Metric Properties of Harmonic Measures

MEMOIRS
of the
American Mathematical Society

Number 867

Metric Properties of Harmonic Measures

Vilmos Totik

November 2006 • Volume 184 • Number 867 (third of 4 numbers) • ISSN 0065-9266

American Mathematical Society
Providence, Rhode Island

2000 *Mathematics Subject Classification.*
Primary 31A15, 31C15, 26C05, 26D05, 30C10, 41A10, 42A05.

Library of Congress Cataloging-in-Publication Data

Totik, V.
Metric properties of harmonic measures / Vilmos Totik.
p. cm. — (Memoirs of the American Mathematical Society, ISSN 0065-9266 ; no. 867)
"Volume 184, number 867 (third of 4 numbers)."
Includes bibliographical references and index.
ISBN-13: 978-0-8218-3994-2 (alk. paper)
1. Harmonic functions. 2. Green's functions. 3. Inequalities (Mathematics) 4. Smoothness of functions. I. Title.

QA405.T68 2006
515′.53—dc22 2006043006

Memoirs of the American Mathematical Society

This journal is devoted entirely to research in pure and applied mathematics.

Subscription information. The 2006 subscription begins with volume 179 and consists of six mailings, each containing one or more numbers. Subscription prices for 2006 are US$624 list, US$499 institutional member. A late charge of 10% of the subscription price will be imposed on orders received from nonmembers after January 1 of the subscription year. Subscribers outside the United States and India must pay a postage surcharge of US$31; subscribers in India must pay a postage surcharge of US$43. Expedited delivery to destinations in North America US$35; elsewhere US$130. Each number may be ordered separately; *please specify number* when ordering an individual number. For prices and titles of recently released numbers, see the New Publications sections of the *Notices of the American Mathematical Society.*

Back number information. For back issues see the *AMS Catalog of Publications.*

Subscriptions and orders should be addressed to the American Mathematical Society, P. O. Box 845904, Boston, MA 02284-5904, USA. *All orders must be accompanied by payment.* Other correspondence should be addressed to 201 Charles Street, Providence, RI 02904-2294, USA.

Memoirs of the American Mathematical Society is published bimonthly (each volume consisting usually of more than one number) by the American Mathematical Society at 201 Charles Street, Providence, RI 02904-2294, USA. Periodicals postage paid at Providence, RI. Postmaster: Send address changes to Memoirs, American Mathematical Society, 201 Charles Street, Providence, RI 02904-2294, USA.

This publication is indexed in *Science Citation Index®*, *SciSearch®*, *Research Alert®*, *CompuMath Citation Index®*, *Current Contents®/Physical, Chemical & Earth Sciences.*
Printed in the United States of America.

♾ The paper used in this book is acid-free and falls within the guidelines established to ensure permanence and durability.
Visit the AMS home page at http://www.ams.org/

10 9 8 7 6 5 4 3 2 1 11 10 09 08 07 06

Contents

Abstract

Smoothness properties of Green functions and harmonic measures on infinitely connected domains $\mathbf{C} \setminus E$ is investigated. The smoothness at a boundary point S is measured in terms of the density function $\Theta_E(t)$ of the circular projection of the boundary onto a half line emanating from S. We shall use the density condition

$$\int \frac{\Theta_E(t)^2}{t^3} dt < \infty, \qquad (1)$$

which will result in optimal order of smoothness. The results are also closely connected with estimates for the distribution of equilibrium measures of compact sets. Disregarding constants, the estimates are sharp, although for sets with some special structure the structure may lead to a better smoothness. In this connection it is proven that for Cantor type sets $\mathcal{C} = \mathcal{C}(\varepsilon_1, \varepsilon_2, \ldots)$ formed with parameters $\{\varepsilon_i\}$ the Green function of $\mathbf{C} \setminus \mathcal{C}$ is of (optimal) Lip 1/2 smooth if and only if $\sum_i \varepsilon_i^2 < \infty$. In particular, this results in a set (say $\mathcal{C}(1/2, 1/3, 1/4, \ldots)$) which is of zero Lebesgue measure but the Green function of its complement is in the Lip 1/2 class.

The conditions can be extended to ensure higher order of smoothness. Applications are given for Phragmén-Lindelöf type theorems on infinitely connected domains; for Markov and Bernstein type polynomial inequalities for general compact sets; for fast decreasing polynomials on compact sets; for Remez and Schur type inequalities and to polynomial approximation of $|x - x_0|^p$ on compact sets. Most of the results have two faces: on the one hand the density condition (1) used in the estimates of harmonic measures and Green functions is sufficient for the classical results to be valid for such more general compact sets. On the other hand, they are also valid for the aforementioned Cantor sets (for which (1) does not hold). It is quite remarkable, that though (1) is not necessary in strict sense, in all of our applications, regarding the density of the set, this condition cannot be weakened.

Received by the editor 2/25/2002.

AMS 2000 Mathematics Subject Classification: 31A15, 31C15, 26C05, 26D05, 30C10, 41A10, 42A05

Key words and phrases: harmonic measures, Green's functions, smoothness, equilibrium measures, circular projection, infinitely connected domains, Cantor sets, Phragmén-Lindelöf theorems, fast decreasing polynomials, polynomial inequalities, polynomial approximation,

1 Introduction

In this paper we are interested in smoothness properties of Green functions and harmonic measures on infinitely connected domains. Several applications will be given in different directions, such as Phragmén-Lindelöf type theorems; Markov and Bernstein type polynomial inequalities; fast decreasing polynomials; Remez and Schur type inequalities and polynomial approximation of $|x-x_0|^p$; all these on general compact sets.

Let E be a compact set on the complex plane, Ω the unbounded component of $\mathbf{C}\setminus E$ and $g_\Omega(z)=g_\Omega(z,\infty)$ the Green function of Ω with pole at infinity (if the Green function does not exist, set g_Ω to be identically ∞). Suppose that 0 is a boundary point of Ω, and our interest is in the smoothness of the Green function g_Ω at 0. We shall measure the density of E at 0 via the circular projection

$$E^c=\{r\,|\,C_r(0)\cap E\neq\emptyset\} \tag{1.1}$$

of E onto the positive real line, where $C_r(a)$ denotes the circle with radius r about the point a. More precisely the density is measured by the function

$$\Theta_E(t)=\Theta(t)=|[0,t]\setminus E^c|,$$

where $|\cdot|$ denotes linear Lebesgue measure. A theorem of Beurling [1, Theorem 3-6] implies that for $x>0$ we have the inequality

$$g_\Omega(-x)\leq g_{\mathbf{C}\setminus E^c}(-x),$$

i.e. among those sets E that have a given circular projection E^c the worst behavior of the Green function occurs for $E=E^c$. Furthermore, if $E^c\subseteq[0,A]$, then

$$g_{\mathbf{C}\setminus E^c}(z)\geq g_{\mathbf{C}\setminus[0,A]}(z)=\log\left|(2z/A-1)+\sqrt{(2z/A-1)^2-1}\right|,$$

which has a $\sqrt{x}$ behavior for $z=-x$, $x>0$ around the origin. Thus, in terms of $\Theta_E(t)$ the best we can hope for the Green function is a Lip 1/2 behavior around the origin. We shall be interested in (possibly best) conditions on Θ that guarantee this optimal behavior.

There is a general estimate for harmonic measures/Green functions due to M. Tsuji [35, Theorem III.67, p. 112]: if E is a subset of the unit disk and E is of positive capacity, then

$$g_\Omega(z)\leq C\exp\left(-\frac{1}{2}\int_{r\in E^c\cap[|z|,1]}\frac{dr}{r}\right),$$

and here integration by parts gives that the right hand side equals (modulo a factor that is bounded away from zero and infinity)

$$\sqrt{|z|}\exp\left(\frac{1}{2}\int_{|z|}^1\frac{\Theta_E(u)}{u^2}du\right).$$

This shows that if

$$\int_0^1\frac{\Theta(u)}{u^2}du<\infty, \tag{1.2}$$

then the Green function is in Lip 1/2 about the origin. We shall see however, that already the condition

$$\int_0^1 \frac{\Theta(u)^2}{u^3} du < \infty \tag{1.3}$$

implies the Lip 1/2 property (recall that $\Theta(u) \leq u$, so (1.3) is a weaker condition than (1.2)), and this is already the best condition in a certain sense. There cannot be a necessary and sufficient condition for the optimal smoothness in terms of the metric measure $\Theta(t)$, since a structure in the set can result in much better estimates for the Green function than what one can get for general sets. In fact, in Chapter 5 we shall characterize those Cantor type sets E for which $g_{\mathbf{C}\setminus E}$ is Lip 1/2 smooth, and as a result we shall get sets $E \subseteq [0,1]$ of linear Lebesgue measure 0 for which $g_{\mathbf{C}\setminus E}$ is Lip 1/2 smooth (for such sets $\Theta_E(u) \equiv u$, hence (1.3) is not satisfied).

There have been several motivations for this work. One was the extension of some well-known polynomial inequalities to general compact sets. T. Erdélyi, A. Kroó and J. Szabados in [18] introduced the density function Θ_E and with it they proved a local version of the Markov inequality (see Chapter 7). Our original motivation was to find the correct form of this local result, and this is where the condition (1.3) emerged. Later it has turned out that the same condition is decisive in many other problems, as is manifested in the different parts of this work. During writing we came across the book [41] by R. K. Vasiliev, where the same condition (1.3) (in a different form) is used in polynomial approximation on general sets. We shall elaborate on Vasiliev's results in Chapter 10.

The main results of this work are about smoothness properties of Green functions and harmonic measures on infinitely connected domains $\mathbf{C}\setminus E$ and about distribution of equilibrium measures of compact sets, as well as several applications of these. Disregarding constants, the estimates will turn out to be sharp, although for sets with some special structure, namely for Cantor-type sets, the structure leads to better results. Applications are given for Phragmén-Lindelöf type theorems on infinitely connected domains; for Markov and Bernstein type polynomial inequalities for general compact sets; for fast decreasing polynomials on compact sets; for Remez and Schur type inequalities and to polynomial approximation of $|x-x_0|^p$ on compact sets.

This work has two faces. On the one hand we shall work with the density function Θ_E and prove results under condition (1.3). In all problems the finiteness of the integral in (1.3) will be sufficient for the given result to hold, while without exception this finiteness turns out to be crucial, i.e. it cannot be replaced by any weaker condition. The other face of this work is about Cantor type sets. Their structure implies stronger results than the ones one can get using the density function Θ_E. If we form a Cantor type set with parameters $\varepsilon_1, \varepsilon_2, \ldots$ (see Chapter 5), then the results that hold under the condition (1.3) will be true if and only if $\sum_i \varepsilon_i^2 < \infty$. Since this allows the set E to have zero linear measure (indeed, such a Cantor set is of zero measure if and only if $\sum_i \varepsilon_i = \infty$), here we witness a family of compact sets for which the main results are true even though the sets can be of zero measure (for such sets (1.3) does not hold).

The original version of this memoir was circulated in manuscript form, and since its writing some papers have emerged that were partially motivated by or are connected with the results in the present work. For smoothness of Green's functions and density of sets (in terms of measure and capacity) see the works [6]–[9] by V. Andrievskii. In particular, [7] generalizes some of the results in Chapter 2 for sets on the real line, while [9] contains an approach to some of the results in this work on Cantor sets which is based on conformal invariants such as moduli of curves. Hölder continuity of Green's functions in terms of capacity condition as well as its relation to Markov inequalities and uniformly perfect sets was treated in [15], and that work also contains a characterization via a Wiener-type series of optimal smoothness (Hölder 1/2 property) of Green's functions $g_{\mathbf{C}\setminus E}$ with $E \subset [0,1]$. See also [33]. For related questions for Newtonian potentials in higher dimension see [34] and the references there, in particular [22]–[25].

The preparation of this work took several years, during which the author was supported by NSF grants DMS-0097484, DMS-0406450 and by OTKA T034323, T049448, TS044782. Part of the work was carried out within the Analysis Research Group of the Hungarian Academy of Sciences.

2 Metric properties of harmonic measures, Green functions and equilibrium measures

Let P be a point on the plane and let $C_r(P)$ resp. $\Delta_r(P)$ denote the circle resp. the open disk with center at P and with radius r.

Suppose that E is a compact set on the plane. We shall measure the density of E at P by the function

$$\Theta_{E,P}(t) = \left|[0,t] \setminus \{r \,|\, C_r(P) \cap E \neq \emptyset\}\right|, \tag{2.1}$$

where $|\cdot|$ denotes linear Lebesgue measure. Thus, the larger E is, the smaller is the function $\Theta_{E,P}$, and a very small $\Theta_{E,P}(t)$ means that for most $r \in [0,t]$ the circle $C_r(P)$ intersects E. In what follows without loss of generality we may assume that E is a subset of the unit disk and P is the origin, in which case we write $\Theta_E(t) = \Theta_{E,0}(t)$.

We recall the definition of harmonic measures. Let G be a domain with bounded boundary and $F \subseteq \partial G$ a closed subset of its boundary. The harmonic measure $\omega(z, F, G)$ is the unique harmonic function in G that lies between 0 and 1 and has boundary values (with the exception of a set of zero capacity) that are equal 1 on F and 0 on $\partial G \setminus F$. In local behavior of harmonic measures we can always assume that G is a domain obtained by removing from a disk a compact set. Exactly as for Green functions, if $G = \Delta_1(0) \setminus E$ where $E \subseteq [0,1]$, $0 \in E$ is a compact set, then for $z = -r$, $r > 0$ we have $\omega(z, C_1(0), \Delta_1(0) \setminus E) \geq c\sqrt{r}$, i.e. such a harmonic measure cannot be smoother than Lip 1/2 about the origin. The first theorem gives an estimate for local smoothness of harmonic measures about a point of the boundary.

Theorem 2.1 *There are absolute constants C, D such that if E is a compact subset of the unit disk of positive capacity and $0 < |z| < 1$, then*

$$\omega(z, C_1(0), \Delta_1(0) \setminus E) \leq C\sqrt{|z|} \exp\left(D \int_{|z|}^1 \frac{\Theta_E^2(u)}{u^3} du\right), \tag{2.2}$$

where Θ_E is the Θ function (2.1) *with respect to E and the origin.*

This implies that for any $0 < r < 1$

$$\omega(0, C_1(0), \Delta_1(0) \setminus E) \leq C\sqrt{r} \exp\left(D \int_r^1 \frac{\Theta_E^2(u)}{u^3} du\right). \tag{2.3}$$

Actually (2.2) and (2.3) are equivalent.

In particular[1], if

$$\int_0^1 \frac{\Theta_E^2(u)}{u^3} du < \infty, \tag{2.4}$$

then the harmonic measure $\omega(z, C_1(0), \Delta_1(0) \setminus E)$ satisfies a Lip 1/2 condition at the origin:

$$\omega(z, C_1(0), \Delta_1(0) \setminus E) \leq C|z|^{1/2}. \tag{2.5}$$

[1] Added in proof: see the paper [15] for a Wiener-type necessary and sufficient capacity condition for (2.5)

Let
$$E^c = \{r \mid C_r(0) \cap E \neq \emptyset\} \tag{2.6}$$
be the circular projection of E onto the positive half line defined in (1.1), and let $\tau_k = 2^k|[2^{-k}, 2^{-k+1}] \setminus E^c|$ be the relative size of the complement of E^c in the interval $[2^{-k}, 2^{-k+1}]$. Then condition (2.4) is equivalent to

$$\sum_k \tau_k^2 < \infty,$$

which shows that in order that (2.4) holds, the circular projection E^c should "cover most of the intervals" $[2^{-k}, 2^{-k+1}]$, $k = 1, 2, \ldots$.

For Green functions the preceding theorem takes the following form.

Theorem 2.2 *There are absolute constants C, D such that if E is a compact subset of the unit disk of positive capacity and $0 < |z| < 1$, then*

$$g_{\mathbf{C}\setminus E}(z) \le C\sqrt{|z|} \exp\left(D \int_{|z|}^1 \frac{\Theta_E^2(u)}{u^3} du\right) \log \frac{2}{\mathrm{cap}(E)}. \tag{2.7}$$

It follows that for any $0 < r < 1$ we have

$$g_{\mathbf{C}\setminus E}(0) \le C\sqrt{r} \exp\left(D \int_{r}^1 \frac{\Theta_E^2(u)}{u^3} du\right) \log \frac{2}{\mathrm{cap}(E)}. \tag{2.8}$$

Actually we shall prove (2.8) first and deduce (2.7) from it.

In particular, if (2.4) holds, then the Green function $g_{\mathbf{C}\setminus E}$ satisfies a local Lip 1/2 condition at the origin:
$$g_{\mathbf{C}\setminus E}(z) \le C|z|^{1/2}. \tag{2.9}$$

We remark (and shall also use) that the factor $\log 1/\mathrm{cap}(E)$ on the right of (2.7) appears only to cover pathological cases (like when $E = [1/2, 1/2 + a]$ with some very small a). In fact, the proof gives the following: Let $\delta > 0$ be fixed. There are constants C, D depending only on δ such that if E is a compact subset of the unit disk, $|z| < 1$, and if there is an s with $|z| \le s$ such that

$$|[s, 2s] \cap E^c| \ge \delta s, \tag{2.10}$$

where E^c is the circular projection of E given in (2.6), then

$$g_{\mathbf{C}\setminus E}(z) \le C\sqrt{|z|} \exp\left(D \int_{|z|}^1 \frac{\Theta_E^2(u)}{u^3} du\right), \tag{2.11}$$

i.e. in this case the factor $\log 1/\mathrm{cap}(E)$ on the right of (2.7) can be omitted.

We shall deduce all these results from the $E \subseteq [0, 1]$ case of the next theorem, where μ_E denotes the equilibrium measure of the set E (see the next section for its definition).

Theorem 2.3 *There are absolute constants C, D such that if E is a compact subset of the unit disk of positive capacity and $0 < r < 1$, then*

$$\mu_E(\Delta_r(0)) \le C\sqrt{r} \exp\left(D \int_{r}^1 \frac{\Theta_E^2(u)}{u^3} du\right). \tag{2.12}$$

In particular, if (2.4) is true, then $\mu_E(\Delta_r(0)) = O(\sqrt{r})$.

Compare this with the fact that $d\mu_{[0,1]}(t) = dt/\pi\sqrt{t(1-t)}$, and hence

$$\mu_{[0,1]}([0,t]) \geq \frac{2}{\pi}\sqrt{r}$$

for all $0 < r < 1$.

The results in this chapter are related to conditions on Hölder continuity of Green's functions, see e.g. [15], and [22]–[25] for the multidimensional case.

2.1 Notations and some basic results from potential theory

In this section we fix our notation, recall some well known theorems from potential theory, and prove some lemmas to be used in later sections. For general references regarding logarithmic potential theory we refer to the books [29], [35], [30], [21], [27].

Let E be a compact subset on the plane and consider its energy

$$V(E) = \inf_{\mu} \int\int \log\frac{1}{|z-t|} d\mu(z)d\mu(t), \tag{2.13}$$

where the infimum is taken for all Borel measures that are supported on E and that have total mass 1. The logarithmic capacity for E is defined as $\mathrm{cap}(E) = e^{-V(E)}$. If E has positive capacity, then there is a unique measure $\mu = \mu_E$, called the equilibrium measure of E, that minimizes the integral (2.13). If we define the logarithmic potential of a measure ν as

$$U^{\nu}(z) = \int \log\frac{1}{|z-t|} d\nu(t), \tag{2.14}$$

then

$$U^{\mu_E}(z) = \log\frac{1}{\mathrm{cap}(E)} \tag{2.15}$$

for all $z \in E$ with the exception of a set of zero capacity, and for all $z \in \mathbf{C}$ we have

$$U^{\mu_E}(z) \leq \log\frac{1}{\mathrm{cap}(E)}.$$

These two properties characterize the equilibrium measures, namely if ν is a measure with total mass 1 and with support in E and if for some constant C we have

$$U^{\nu}(z) = C$$

for all $z \in E$ with the exception of a set of zero capacity, and

$$U^{\nu}(z) \leq C, \qquad \text{for all } z \in \mathbf{C},$$

then $\nu = \mu_E$. The set E is called a regular set if (2.15) holds for all $z \in E$.

If ν is any measure supported on E and of total mass 1, then

$$\min_{z\in E} U^{\nu}(z) \leq \log\frac{1}{\mathrm{cap}(E)} \leq \max_{z\in E} U^{\nu}(z). \tag{2.16}$$

In an important special case we have a fairly explicit form for the equilibrium measure: Let $E = \cup_{j=1}^{m}[a_j, b_j]$, $a_1 < b_1 < a_2 < b_2 \cdots b_{m-1} < a_m < b_m$ consist of finitely many intervals. In this case (see e.g. [31, Lemma 4.4.1]) μ_E is absolutely continuous with respect to linear Lebesgue measure, and $d\mu_E(x) = \omega_E(x)dx$, with

$$\omega_E(x) = \frac{\prod_{j=1}^{m-1} |x - \lambda_j|}{\pi\sqrt{\prod_{j=1}^{m} |x - a_j||x - b_j|}}, \tag{2.17}$$

where λ_j are chosen so that

$$\int_{b_k}^{a_{k+1}} \frac{\prod_{j=1}^{m-1} (t - \lambda_j)}{\sqrt{\prod_{j=1}^{m} |t - a_j||t - b_j|}} dt = 0$$

for all $k = 1, \ldots, m-1$.

The Green function $g_{\mathbf{C}\setminus E}(z)$ of $\mathbf{C} \setminus E$ is defined as the Green function $g_\Omega(z, \infty)$ with pole at infinity, where Ω denotes the unbounded component of $\mathbf{C} \setminus E$. This Green function exists if and only if E is of positive capacity and for E with positive capacity it is the unique function g that satisfies the following conditions:

- $g \geq 0$, g is harmonic in Ω,
- $g(z) = \log |z| + O(1)$ as $z \to \infty$, $z \in \Omega$,
- $\lim_{w\to z,\ w\in\Omega} g(w) = 0$ for all $z \in \partial\Omega$ with the exception of a set of zero capacity.

Often we also define $g_{\mathbf{C}\setminus E}(z) = 0$ for $z \notin \Omega$, by which it becomes a subharmonic function.

In turns of the equilibrium measure we have the representation

$$g_{\mathbf{C}\setminus E}(z) = \log \frac{1}{\operatorname{cap}(E)} - U^{\mu_E}(z). \tag{2.18}$$

Thus, in terms of the Green functions the regularity of E means that the Green function $g_{\mathbf{C}\setminus E}$ is continuous on the whole plane. Hence, the last condition in the definition of the Green function is true for all boundary point z of Ω. Instead of saying that E is a regular set it is more customary to say that Ω is regular with respect to the Dirichlet problem. Thus E is regular if and only if every boundary point z of Ω is a regular point in the sense that

$$\lim_{w\to z,\ w\in\Omega} g_\Omega(w) = 0.$$

By a theorem of Wiener this local regularity property can be characterized as follows: Let $0 < \lambda < 1$ and set

$$A_n(z) := \left\{y \,\middle|\, y \notin G,\ \ \lambda^n \leq |y - z| < \lambda^{n-1}\right\}.$$

Then $z \in \partial G$ is a regular boundary point of Ω if and only if

$$\sum_{n=1}^{\infty} \frac{n}{\log(1/\operatorname{cap}(A_n(z))} = \infty. \tag{2.19}$$

In particular, if ever boundary point is the endpoint of an arc lying in E, then E is regular.

One application of the Green function is the so called Bernstein–Walsh lemma ([42, p. 77])

$$|P_n(z)| \le e^{ng_{\mathbf{C}\setminus E}(z)}\|P_n\|_E, \tag{2.20}$$

which estimates the size of a polynomial of degree at most n at an arbitrary point of the plane in terms of its supremum norm on the compact set E. It is also true that this is the best estimate in the sense that

$$\sup \left(\frac{|P_n(z)|}{\|P_n\|_E}\right)^{1/n} = e^{g_{\mathbf{C}\setminus E}(z)}, \tag{2.21}$$

where the supremum is taken for all polynomials P_n of degree $n = 1, 2, \ldots$. When $E = [-1, 1]$, then (2.20) takes the form

$$|P_n(z)| \le |z + \sqrt{z^2 - 1}|^n \|P_n\|_{[-1,1]}, \tag{2.22}$$

where that branch of $\sqrt{z}$ is taken which is positive for positive z. Now this implies

$$|P_n(z)| \le (2|z|)^n \|P_n\|_{[-1,1]}, \tag{2.23}$$

or the equivalent form: if $J \subseteq E$ is an interval, then

$$|P_n(z)| \le \left(\frac{4\mathrm{diam}(z, J)}{|J|}\right)^n \|P_n\|_E. \tag{2.24}$$

In this work we shall extensively use the notion of balayage (or sweeping out). If ν is a measure and E is a compact set of positive capacity, then the balayage of ν onto E is defined as the unique measure $\widehat{\nu}$ that is supported on E, has the same total mass as ν, and for which $U^{\widehat{\nu}}(z) = U^\nu(z) + c$ is true for all $z \in E$ with the exception of a set of capacity zero. Here both $\widehat{\nu}$ and c are uniquely determined by these properties, and actually c is equal to $\int_\Omega g_\Omega(z)d\nu(z)$, where Ω denotes the unbounded component of $\mathbf{C} \setminus E$. We have for all $z \in \mathbf{C}$ the inequality $U^{\widehat{\nu}}(z) \le U^\nu(z) + c$, and if Ω is a regular domain for the Dirichlet problem, then on E we have equality:

$$U^{\widehat{\nu}}(z) = U^\nu(z) + c \qquad z \in E.$$

For taking balayage onto a set consisting of finitely many intervals we have ([38, Lemma 2.3]): let $\Sigma = \cup_{i=1}^l [a_i, b_i]$, $(b_l, a_{l+1}) = (b_l, \infty] \cup (-\infty, a_1)$, and for $a \in \overline{\mathbf{R}} \setminus \Sigma$ let $i(a)$ be that index $1 \le i \le l$ for which $a \in (b_i, a_{i+1})$. The density of the balayage measure $\widehat{\delta_a}$ of the Dirac delta mass δ_a onto Σ is given by

$$\frac{1}{\pi} \frac{\prod_1^l |(a - a_j)(a - b_j)|^{1/2}}{\prod_1^l |(t - a_j)(t - b_j)|^{1/2}} \frac{|P_{l-1}(t)|}{|P_{l-1}(a)|} \frac{1}{|t - a|}, \qquad t \in \Sigma, \tag{2.25}$$

where the polynomial

$$P_{l-1}(t) = \prod_{1 \le i \le l,\ i \ne i(a)} (t - \tau_i)$$

satisfies for all $1 \le i \le l$, $i \ne i(a)$ the condition

$$\int_{b_i}^{a_{i+1}} \frac{\prod_1^l |(a - a_j)(a - b_j)|^{1/2}}{\prod_1^l |(t - a_j)(t - b_j)|^{1/2}} \frac{P_{l-1}(t)}{P_{l-1}(a)} \frac{1}{|t - a|} dt = 0. \tag{2.26}$$

This system of equations uniquely determines each τ_i, $i \neq i(a)$, and we have $\tau_i \in (b_i, a_{i+1})$.

The equilibrium measure μ_Σ can be regarded as the balayage of δ_∞ onto Σ (see below). In particular, for $a \to \infty$ we get from the preceding formulae that the equilibrium measure μ_Σ is of the form 2.17).

If $E^* \subseteq E$ are sets of positive capacity, then for their equilibrium measures we have $\mu_{E^*} = \widehat{\mu_E}$, where $\widehat{\nu}$ denotes the balayage of ν onto E^*. In particular, for $F \subseteq E^*$ we have $\mu_E(F) \leq \mu_{E^*}(F)$. Also, the procedure of taking balayage onto E^* can be obtained in two stages: first taking balayage onto E, and then taking balayage onto E^*. The balayage of the normalized arc measure on a circle C onto a set E lying inside C is the equilibrium measure μ_E on E. By blowing up C we can see that in a certain way the equilibrium measure μ_E is the balayage of the Dirac mass δ_∞ at infinity.

Let G be a domain with bounded boundary and $F \subseteq \partial G$ a closed subset of its boundary. The harmonic measure $\omega(z, F, G)$ is the unique harmonic function in G that has values in $[0,1]$ and has boundary values (with the exception of a set of zero capacity) that are equal 1 on F and 0 on $\partial G \setminus F$. This $\omega(z, E, G)$ can be expressed as $\omega(z, E, G) = \widehat{\delta_z}(E)$, where $\widehat{\delta_z}$ denotes the balayage of the Dirac delta mass δ_z at $z \in G$ onto the boundary ∂G of G. If $G = \mathbf{C} \setminus [\alpha, \beta]$, i.e. $\partial G = [\alpha, \beta]$, then (2.25) takes the following form for the balayage measure $\widehat{\delta_a}$, $a \in \mathbf{R} \setminus [\alpha, \beta]$, onto $[\alpha, \beta]$ (see also [30, Ch. II., (4.47)]):

$$d\widehat{\delta_a}(t) = \frac{1}{\pi} \frac{|\sqrt{(a-\alpha)(a-\beta)}|}{|t-a|\sqrt{(t-\alpha)(\beta-t)}} dt. \tag{2.27}$$

Thus, if ν is a measure supported on $\mathbf{R} \setminus [\alpha, \beta]$, then the balayage measure onto $[\alpha, \beta]$ is absolutely continuous on $[\alpha, \beta]$, and for $t \in (\alpha, \beta)$ we have

$$\frac{d\widehat{\nu}(t)}{dt} = \frac{1}{\pi} \int \frac{|\sqrt{(a-\alpha)(a-\beta)}|}{|t-a|\sqrt{(t-\alpha)(\beta-t)}} d\nu(a). \tag{2.28}$$

We need one more fact regarding balayage measures. Let $F \subset \mathbf{R}$ be a compact subset of the real line and let $I \subset F$ be an open interval inside F. If F is an interval, then for any $a \in \mathbf{C} \setminus F$ the balayage of δ_a onto F is given on I by ([30, (Ch. II, (4.40)])

$$\frac{d\widehat{\delta_a}(t)}{dt} = \frac{1}{2\pi} \left(\frac{\partial g_{\mathbf{C}\setminus F}(t,a)}{\partial \mathbf{n}_+} + \frac{\partial g_{\mathbf{C}\setminus F}(t,a)}{\partial \mathbf{n}_-} \right),$$

where $\mathbf{n}_\pm$ denote the upper and lower normals to F at $t \in I$, and $g_{\mathbf{C}\setminus F}(x,a)$ is the Green function of $\mathbf{C}\setminus F$ with pole at $a \in \mathbf{C}\setminus F$. Thus, on $I \subset F$ the balayage measure $\widehat{\delta_a}$ is absolutely continuous with respect to Lebesgue measure, its density $d\widehat{\delta_a}(t)/dt$ is continuous (actually a C^∞ function), and $d\widehat{\delta_a}(t)/dt$ is a harmonic function of $a \in \mathbf{C} \setminus F$. Now these properties all hold true if $F \subset \mathbf{R}$ is any compact set and $I \subset \mathbf{R}$ is an open interval inside F. In fact, if $\widehat{\nu}$ denotes the balayage onto F and $\overline{\nu}$ denotes the balayage onto $\overline{I}$, then

$$\overline{\delta_a} = \widehat{\delta_a}\Big|_{\overline{I}} + \overline{\widehat{\delta_a}\Big|_{F\setminus\overline{I}}},$$

and here, as we have just mentioned (c.f. also (2.28)), both $\overline{\delta_a}$ and

$$\overline{\widehat{\delta_a}\Big|_{F\setminus\overline{I}}} = \int_{F\setminus\overline{I}} \overline{\delta_b}\, d\widehat{\delta_a}(b)$$

are absolutely continuous on I with continuous density that depends harmonically on $a \in \mathbf{C}\setminus F$, and so the same is true of their difference $\widehat{\delta_a}\Big|_I$.

We also recall Harnack's inequality for nonnegative harmonic functions $w(z)$ in the unit disk: for all $|z|<1$ we have

$$\frac{1-|z|}{1+|z|}w(0) \le w(z) \le \frac{1+|z|}{1-|z|}w(0). \tag{2.29}$$

In particular, for $|z| \le 1/2$

$$\frac{1}{3}w(0) \le w(z) \le 3w(0). \tag{2.30}$$

This yields the following: if $E \subset \mathbf{R}$ is a compact set and $w(z)$ is a nonnegative harmonic function on $\overline{\mathbf{C}}\setminus E$, then for all $z \in \mathbf{C}$ with $\mathrm{dist}(z,E) \ge \mathrm{diam}(E)/2$ we have

$$\frac{1}{3}w(\infty) \le w(z) \le 3w(\infty). \tag{2.31}$$

In fact, if I denotes the smallest interval that contains E, then w is harmonic outside I, and we can use a conformal mapping from $\mathbf{C}\setminus I$ onto the unit disk to transform the problem to (2.30).

2.2 Preliminary estimates

In what follows we shall need some elementary estimates for harmonic measures and balayage measures.

The proof of the following lemma is an easy adaptation of the proof of a theorem of Beurling (see [27, Sec. IV.5.4]). It says that for symmetric sets $F \subset [-1,1]$ of given Lebesgue measure the harmonic measure $\omega(0, C_1(0), \Delta_1(0)\setminus F)$ is the largest when F consists of two intervals that are as far from the origin as possible.

Lemma 2.4 *Let $F \subset [-1,1]$ be closed and symmetric with respect to the origin, and let $|[-1,1]\setminus F| = 2\tau$. Then*

$$\omega(0, C_1(0), \Delta_1(0)\setminus F) \le \omega(0, C_1(0), \Delta_1(0)\setminus([-1,-\tau]\cup[\tau,1])). \tag{2.32}$$

Furthermore, if $(-a,a)\cap F = \emptyset$, then $\omega(x, C_1(0), \Delta_1(0)\setminus F)$ is increasing on the interval $(-a,0)$ and decreasing on $(0,a)$.

Proof. Without loss of generality we may assume that F consists of a finite number of intervals.

Let

$$\omega(z) = 1-\omega(z, C_1(0), \Delta_1(0)\setminus F) = \omega(z, F, \Delta_1(0)\setminus F).. \tag{2.33}$$

By [27, (5.4), p.105] we have

$$\omega(z) = \frac{1}{2\pi}\int_F g(\xi,z)\,d\tilde{\omega}(\xi),$$

where $\tilde{\omega}$ is the harmonic conjugate of the function ω with $\tilde{\omega}(0) = 0$, the integral is taken on both parts of the cut $\mathbf{C} \setminus F$, and

$$g(\xi, z) = \log \left| \frac{1 - \bar{\xi} z}{\xi - z} \right|$$

is the Green function of $\Delta_1(0)$ with pole at z. By symmetry,

$$\omega(z) = \frac{1}{\pi} \int_F g(\xi, z) d\tilde{\omega}(\xi),$$

where the integral is taken on the upper side of the cut. For $\xi \in F$ we have

$$\left. \frac{\partial \omega(z)}{\partial y} \right|_{z = \xi} \leq 0,$$

hence $d\tilde{\omega} \geq 0$ on F follows from the Cauchy-Riemann equations. In other words, the preceding integral is against a positive measure.

Since the set F is symmetric with respect to the origin, the function ω is symmetric with respect to the imaginary axis, and hence $d\tilde{\omega}(t)/dt$ is an even function on F by the Cauchy-Riemann equations. Therefore, we have

$$\omega(x) = \frac{1}{\pi} \int_{F \cap [0,1]} [g(\xi, x) + g(-\xi, x)] \, d\tilde{\omega}(\xi). \tag{2.34}$$

Here

$$g(\xi, x) + g(-\xi, x) = \log \left| \frac{1 - \xi^2 x^2}{\xi^2 - x^2} \right|,$$

the derivative of which with respect to x is

$$\frac{d}{dx}(g(\xi, x) + g(-\xi, x)) = 2x \frac{-1 + \xi^4}{(1 - \xi^2 x^2)(x^2 - \xi^2)} \begin{cases} \geq 0 & \text{if } 0 \leq x \leq \xi \\ \leq 0 & \text{if } \xi < x < 1. \end{cases} \tag{2.35}$$

In view of (2.33) and (2.34) this verifies the last statement of the lemma.

We have assumed that F consists of finitely many intervals, say

$$F \cap [0, 1] = [a_1, b_1] \cup [a_2, b_2] \cup \cdots \cup [a_n, b_n],$$

where

$$0 \leq a_1 < b_1 < a_2 < b_2 \cdots < a_n < b_n \leq 1.$$

First we show that if F_1 is symmetric onto the origin and

$$F_1 \cap [0, 1] = [a_2 - (b_1 - a_1), b_2] \cup [a_3, b_3] \cup \cdots \cup [a_n, b_n],$$

then

$$\omega(0, F, \Delta_1(0) \setminus F) \geq \omega(0, F_1, \Delta_1(0) \setminus F_1) \tag{2.36}$$

(note that $F_1 \cap [0, 1]$ is obtained from $F \cap [0, 1]$ by shifting the leftmost interval $[a_1, b_1]$ to the right until it reaches $[a_2, b_2]$). In fact, let

$$\omega(z) = \omega_1(z) + \omega_2(z),$$

where

$$\omega_1(x) = \frac{1}{\pi} \int_{[a_1, b_1]} [g(\xi, x) + g(-\xi, x)] \, d\tilde{\omega}(\xi)$$

and

$$\omega_2(x) = \frac{1}{\pi}\sum_{j=2}^{n}\int_{[a_j,b_j]}[g(\xi,x)+g(-\xi,x)]\,d\tilde{\omega}(\xi).$$

The function

$$v(z) = \omega_1\left(\frac{b_1}{a_2}z\right) + \omega_2(z)$$

is harmonic outside the set symmetric set F_1' for which

$$F_1' \cap [0,1] = \left[\frac{a_2}{b_1}a_1, a_2\right] \cup [a_2,b_2] \cup \cdots \cup [a_n,b_n].$$

Since $b_1 - a_1 \leq a_2(b_1-a_1)/b_1$, we get

$$\frac{a_2}{b_1}a_1 \leq a_2 - (b_1 - a_1),$$

which shows that $F_1 \subseteq F_1'$, and hence

$$\omega(0, F_1', \Delta_1(0)\setminus F_1') \geq \omega(0, F_1, \Delta_1(0)\setminus F_1).$$

Hence, to complete the proof of (2.36) it is enough to verify that

$$\omega(0,F,\Delta_1(0)\setminus F) = v(0) \geq \omega(0,F_1',\Delta_1(0)\setminus F_1'),$$

and this follows from the maximum principle if we can show that for $x \in F_1'$ we have $v(x) \geq 1$. If $x \in [a_2a_1/b_1, a_2]$, then (2.35) yields

$$v(x) \geq \omega_1\left(\frac{b_1}{a_2}x\right) + \omega_2\left(\frac{b_1}{a_2}x\right) = \omega\left(\frac{b_1}{a_2}x\right) = 1.$$

If, however, $x \in \cup_{j=2}^{n}[a_j,b_j]$, then we obtain again from (2.35) that

$$v(x) \geq \omega_1(x) + \omega_2(x) = \omega(x) = 1,$$

and this completes the proof of (2.36).

Now $F_1 \cap [0,1]$ consists of $(n-1)$ intervals (we have pushed the first interval $[a_1,b_1]$ until we hit the next one), and in the process of going from F to F_1 the value of $\omega(0,F,\Delta_1(0)\setminus F)$ decreases. Repeating this process $(n-2)$ more times we arrive at a set $F_{n-1} = [-b,-a]\cup[a,b]$ with $b-a = 1-\tau$, and

$$\omega(0,F,\Delta_1(0)\setminus F) \geq \omega(0,F_{n-1},\Delta_1(0)\setminus F_{n-1}).$$

Now looking at boundary values on $[a/b,1]$ we can see that

$$\omega(bz,F_{n-1},\Delta_1(0)\setminus F_{n-1}) \geq \omega\left(z,[-1,-\frac{a}{b}]\cup[\frac{a}{b},1],\Delta_1(0)\setminus([-1,-\frac{a}{b}]\cup[\frac{a}{b},1])\right),$$

and since here $[-1,-a/b]\cup[a/b,1]$ contains $[-1,-\tau]\cup[\tau,1]$, the claim of the lemma follows.

■

Lemma 2.5 *Let $F \subset [-1,1]$ be closed and let $\tau = |[-1,1]\setminus F|$. Then*

$$\omega(0, C_1(0), \Delta_1(0)\setminus F) \leq \frac{4}{\pi}\tau. \tag{2.37}$$

Proof. Let $F^* = F \cap (-F)$. This is symmetric with respect to the origin, and $|[-1,1] \setminus F^*| \le 2\tau$. By Lemma 2.4 we can see that

$$\begin{aligned}\omega(0, C_1(0), \Delta_1(0) \setminus F) &\le \omega(0, C_1(0), \Delta_1(0) \setminus F^*)\\ &\le \omega(0, C_1(0), \Delta_1(0) \setminus ([-1,-\tau] \cup [\tau, 1])).\end{aligned}$$

But

$$\omega(z, C_1(0), \Delta_1(0) \setminus ([-1,-\tau] \cup [\tau, 1])) = \omega(z^2, C_1(0), \Delta_1(0) \setminus [\tau^2, 1])),$$

and for the latter value we get from [27, (5.12), p. 111] that it equals

$$\begin{aligned}1 \quad &- \quad \frac{2}{\pi} \arcsin \frac{1-\tau^2}{1+\tau^2} = \frac{2}{\pi} \int_{(1-\tau^2)/(1+\tau^2)}^{1} \frac{1}{\sqrt{1-u^2}} du\\ &\le \quad \frac{2}{\pi\sqrt{2/(1+\tau^2)}} \int_{(1-\tau^2)/(1+\tau^2)}^{1} \frac{1}{\sqrt{1-u}} du = \frac{2}{\pi} \frac{\sqrt{1+\tau^2}}{\sqrt{2}} 2\sqrt{\frac{2\tau^2}{1+\tau^2}} = \frac{4}{\pi}\tau,\end{aligned}$$

and this completes the proof. ∎

Lemma 2.6 *Let $0 < \alpha < 1$, F a closed set, $[0,1] \subset F$ and suppose that*

$$|[1, 2+\alpha] \setminus F| \le \tau.$$

If $\widehat{\delta_a}$ denotes the balayage of the Dirac delta mass δ_a onto F, then for $a \in [1,2]$ we have the estimate

$$\frac{d\widehat{\delta_a}(t)}{dt} \le B\frac{\tau}{\sqrt{t}}, \qquad 0 < t < 1-\alpha, \tag{2.38}$$

where the constant $B = B_\alpha$ depends only on α.

The proof gives that if $\alpha = 1/26$, then $B_\alpha \le 3400$.

We shall actually use this lemma in a scaled form, namely let F be a closed set, $s > 0$, $[0, s] \subset F$ and suppose that

$$|[s, (2+\alpha)s] \setminus F| \le \tau. \tag{2.39}$$

If $\widehat{\delta_a}$ denotes the balayage of δ_a onto F, then for $a \in [s, 2s]$ we have the estimate

$$\frac{d\widehat{\delta_a}(t)}{dt} \le B\frac{\tau}{\sqrt{t}}\frac{1}{s^{3/2}}, \qquad 0 < t < (1-\alpha)s, \tag{2.40}$$

where the constant $B = B_\alpha$ is the same is in the lemma.

We shall also need that the proof gives that for $\alpha = 1/26$ the inequality (2.40) is already true if instead of (2.39) we require only that

$$[a - s/39, a + s/39] \setminus F| \le \tau. \tag{2.41}$$

Proof. To get some concrete estimates to be used in the proof of the main theorems, we shall carry out the proof only for $\alpha = 1/26$. This is the case that will be used, and the proof for other α's is the same. It is again enough to prove the lemma for F consisting of finitely many intervals.

Thus, let $\alpha = 1/26$, and set $q = 38/39$. First of all (2.27) gives that if $\overline{\nu}$ denotes the balayage onto $[0, q]$, then for any $a > 1$ and $t < 1 - \alpha$

$$\begin{aligned} \frac{d\overline{\delta_a}(t)}{dt} &= \frac{1}{\pi}\frac{\sqrt{a(a-q)}}{(a-t)\sqrt{t(q-t)}} \\ &< \frac{1}{\pi}\frac{\sqrt{1(1-q)}}{(1-(1-\alpha))\sqrt{t(q-(1-\alpha))}} = \frac{26\sqrt{2}}{\pi}\frac{1}{\sqrt{t}}. \end{aligned}$$

Here $d\overline{\delta_a}(t)/dt$ on the left hand side is harmonic (in a) outside $[0, q]$, and the distance of a and $[0, q]$ satisfies $\operatorname{dist}(a, [0, q]) \geq 1 - q = 1/39$, so we get from Harnack's inequality (2.29) that for $|z - a| \leq \rho/39$, $\rho < 1$ the inequality

$$\frac{d\overline{\delta_z}(t)}{dt} \leq \frac{26\sqrt{2}}{\pi}\frac{1+\rho}{1-\rho}\frac{1}{\sqrt{t}}, \qquad t \in (0, 1-\alpha) \tag{2.42}$$

holds. Now we use that $[0, q] \subset F$, hence $\overline{\delta_z} \geq \widehat{\delta_z}\Big|_{[0,q]}$. Therefore (2.42) implies for all $|z - a| = \rho/39$ the same inequality with $d\overline{\delta_z}(t)/dt$ replaced by $d\widehat{\delta_z}(t)/dt$. Since for $t \in (0, 1-\alpha)$ the function $z \to d\widehat{\delta_z}(t)/dt$ is harmonic outside F and vanishes on F, it follows that

$$\frac{d\widehat{\delta_a}(t)}{dt} \leq \frac{26\sqrt{2}}{\pi}\frac{1+\rho}{1-\rho}\frac{1}{\sqrt{t}}\omega(a, C_{\rho/39}(a), \Delta_{\rho/39}(a) \setminus F).$$

The last factor is the same as $\omega(0, C_1(0), \Delta_1(0) \setminus F')$, where $F' = (F - a)39/\rho$. By the assumption on F we have $[-1, 1] \setminus F' \leq 39\tau/\rho$, and so it follows from Lemma 2.5 that

$$\frac{d\widehat{\delta_a}(t)}{dt} \leq \frac{26\sqrt{2}}{\pi}\frac{1+\rho}{1-\rho}\frac{1}{\sqrt{t}}\frac{4}{\pi}\frac{39\tau}{\rho},$$

and with $\rho = \sqrt{2} - 1$ this gives

$$\frac{d\widehat{\delta_a}(t)}{dt} \leq \frac{3400}{\sqrt{t}}\tau.$$

The estimate (2.40) is just a scaled version of the lemma. In fact, if $F^* = F/s$, then $|[0, 2+\alpha] \setminus F^*| \leq \tau/s$, so if $\widetilde{\delta_{a/s}}$ denotes the balayage of $\delta_{a/s}$ onto F^*, then the lemma implies for $x = t/s \in (0, 1-\alpha)$ that

$$d\widehat{\delta_a}(t) = d\widetilde{\delta_{a/s}}(x) \leq B\frac{\tau}{s\sqrt{x}}dx = B\frac{\tau}{s\sqrt{t/s}}\frac{dt}{s} = B\frac{\tau}{\sqrt{t}}\frac{1}{s^{3/2}}dt.$$

∎

2.3 Proof of Theorem 2.3 for $E \subseteq [0,1]$

In this section we prove Theorem 2.3 in the special case when the set E lies on $[0,1]$. Thus, in what follows we assume $E \subseteq [0,1]$, and let $\Theta(t) = \Theta_E(t) = |[0,t] \setminus E|$.

For the convenience of the reader we summarize the values of the different constants B_i used below and indicate their choice in the proof:

$$B_1 = 3400, \qquad B_2 = 5 \cdot 10^4 \geq B_1(1 + 4(2+\sqrt{2})),$$

$$B_3 = 700 \geq \frac{\sqrt{26}}{\pi} + \frac{4}{7} + \frac{\sqrt{2}}{7} B_1,$$

$$B_4 = 2 \cdot 10^4 \geq B_3 \sqrt{\frac{52}{25}} \left(2 + \frac{4(4+2/26)(1+\sqrt{2})}{3} + \frac{1}{7} \right).$$

Let

$$d\mu_1(x) = \frac{1}{\pi\sqrt{x(1-x)}} dx$$

be the equilibrium measure of $[0,1]$, and μ_k the equilibrium measure of the set $E_k = E \cup [0, 2^{-k+1}]$. We set $\alpha = 1/26$, and let S_k be the smallest constant for which

$$d\mu_k(t) \leq \frac{S_k}{\sqrt{t}} dt \qquad \text{for} \quad t \in \left(0, (1-\alpha)2^{-k+1}\right).$$

We shall estimate this S_k in a recursive manner. Clearly,

$$S_1 = \sup_{0<t<1-\alpha} \frac{1}{\pi\sqrt{1-t}} \leq \frac{\sqrt{26}}{\pi}. \tag{2.43}$$

We set $I_k = [2^{-k}, 2^{-k+1}]$ and $J_k = [(1-\alpha)2^{-k+1}, 2^{-k+1}]$ (see Figure 1). Thus, J_k is the rightmost 2α portion of I_k. In the estimation of S_k we shall also need the size of

$$m_k := \mu_k(J_k \setminus E) = \mu_k\left([(1-\alpha)2^{-k+1}, 2^{-k+1}] \setminus E\right).$$

The measure μ_{k+1} is the balayage of μ_k onto E_{k+1}, which means that we obtain μ_{k+1} by adding to $\mu_k|_{E_{k+1}}$ the balayage of $\mu_k|_{I_k \setminus E}$ onto E_{k+1}. For $k \geq 2$ the set E_{k+1} contains $F_k := [0, 2^{-k}] \cup (E \cap [0, (2+\alpha)2^{-k}])$, hence on $[0, 2^{-k}]$ the balayage mentioned last is not bigger than the balayage of $\mu_k|_{I_k \setminus E}$ onto F_k. Since with $s = 2^{-k}$ we have $|[s, s(2+\alpha)] \setminus F_k| \leq \Theta((2+\alpha)2^{-k})$, it follows from (2.40) that for any $a \in I_k \setminus E$ the density of the balayage of δ_a onto F_k is at most

$$\frac{B_1}{\sqrt{t}} \Theta((2+\alpha)2^k) \left(2^k\right)^{3/2} \qquad \text{for} \quad t \in (0, (1-\alpha)2^{-k}) \tag{2.44}$$

with $B_1 = 3400$. The total mass of μ_k on $I_k \setminus E$ is at most

$$\mu_k((I_k \setminus J_k) \setminus E) + \mu_k(J_k \setminus E) \leq \frac{S_k}{\sqrt{2^{-k}}} \Theta(2^{-k+1}) + m_k$$

by the choice of the numbers S_k and m_k. Thus,

$$S_{k+1} \leq S_k + B_1 \Theta((2+\alpha)2^{-k}) \left(2^k\right)^{3/2} \left(S_k \Theta(2^{-k+1}) \sqrt{2^k} + m_k \right),$$

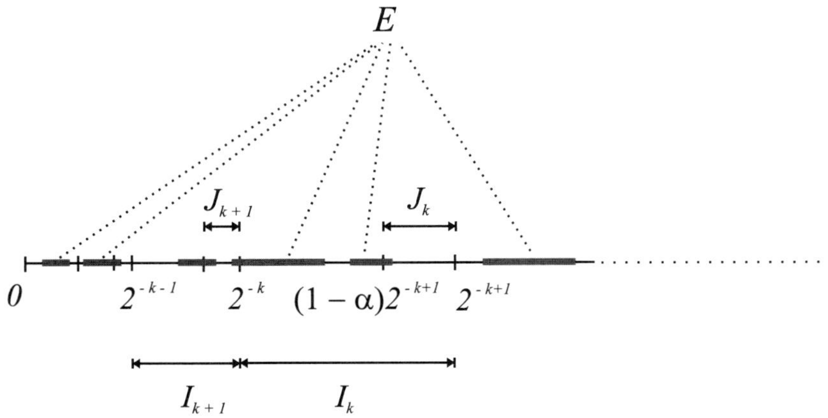

Figure 1:

i.e. if for $k \geq 2$ we set $\theta_k = 2^k\Theta((2+\alpha)2^{-k})$ (θ_1 will have a different meaning), then it follows that

$$S_{k+1} \leq S_k(1+B_1\theta_k^2) + B_1 m_k \theta_k 2^{k/2}. \tag{2.45}$$

This derivation is valid for all $k \geq 2$.

For $k = 1$ (2.45) changes somewhat. The reasoning is valid for the part of μ_1 lying in $[1/2, 1-\alpha] \setminus E$ (see the remark made after Lemma 2.6), in particular (2.44) holds true for $a \in [1/2, (1-\alpha)] \setminus E$ with $\Theta((2+\alpha)2^k)$ replaced by $\theta_1/2 := \Theta(1 - 1/26 + 1/2 \cdot 39) = \Theta(38/39)$ (see (2.41) with $s = 1/2$). However, for $a \in [1-\alpha, 1] \setminus E$ we cannot use Lemma 2.6. In that case we get that on $(0, (1-\alpha)/2]$ the density of the balayage of δ_a onto E_2 is at most as large as the density of the balayage onto $[0, 1/2]$, which, using (2.27), is at most as large as

$$\frac{1}{\sqrt{t}}\frac{1}{\pi}\frac{\sqrt{a(a-1/2)}}{(a-t)\sqrt{1/2-t}} \leq \frac{1}{\sqrt{t}}\frac{1}{\pi}\frac{\sqrt{(1-\alpha)((1-\alpha)-1/2)}}{((1-\alpha)-(1-\alpha)/2)\sqrt{1/2-(1-\alpha)/2}} \leq \frac{4}{\sqrt{t}}.$$

Thus, exactly as above we obtain

$$S_2 \leq S_1(1+B_1\theta_1^2) + 4m_1,$$

and here

$$m_1 \leq \frac{1}{\pi}\int_{1-\alpha}^{1}\frac{dx}{\sqrt{x(1-x)}} \leq \frac{1}{\pi\sqrt{1-\alpha}}\int_{1-\alpha}^{1}\frac{dx}{\sqrt{1-x}} \leq \frac{1}{7}. \tag{2.46}$$

Thus, for $k = 1$ we can write (c.f. (2.43))

$$S_2 \leq \frac{\sqrt{26}}{\pi}(1+B_1\theta_1^2) + \frac{4}{7}, \tag{2.47}$$

where now $\theta_1 = 2\Theta(38/39)$.

Next we estimate m_{k+1}. This is equal to $\mu_k(J_{k+1} \setminus E)$ plus what measure the set $J_{k+1} \setminus E$ gets in taking the balayage of $\mu_k\big|_{I_k \setminus E}$ onto E_{k+1}. This latter measure has two parts, namely the measure of $J_{k+1}\setminus E$ when sweep out $\mu_k\big|_{J_k\setminus E}$ and $\mu_k\big|_{(I_k\setminus J_k)\setminus E}$, respectively, onto E_{k+1}.

If we take the balayage of $\mu_k\big|_{J_k\setminus E}$ onto E_{k+1}, then on $(0,2^{-k})$ the density of the balayage measure is smaller than the density when we take the balayage onto $[0,2^{-k}]$. Therefore, in view of (2.27) the contribution of the balayage of $\mu_k\big|_{J_k\setminus E}$ to m_{k+1} is at most

$$\int_{J_k\setminus E}\int_{(1-\alpha)2^{-k}}^{2^{-k}}\frac{\sqrt{a(a-2^{-k})}}{\pi(a-t)\sqrt{t(2^{-k}-t)}}dtd\mu_k(a)\le\frac{1}{4\sqrt{2}}m_k. \tag{2.48}$$

This is a crucial step in the proof, and it follows from the fact that for $a\in J_k$, $a=2^{-k}b$ where $b\in[2(1-\alpha),2]$ the inner integral is the same as

$$\int_{1-\alpha}^{1}\frac{\sqrt{b(b-1)}}{\pi(b-t)\sqrt{t(1-t)}}dt\le\int_{1-\alpha}^{1}\frac{\sqrt{(2-2\alpha)(1-2\alpha)}}{\pi(2-2\alpha-t)\sqrt{t(1-t)}}dt\le\frac{1}{4\sqrt{2}}$$

by the choice $\alpha=1/26$ (α was chosen exactly so that we have this inequality). Using this, (2.48) is immediate since $\mu_k(J_k\setminus E)=m_k$.

The second contribution to m_{k+1} coming from sweeping out $\mu_k\big|_{(I_k\setminus J_k)\setminus E}$ onto E_{k+1} is at most the total mass $\mu_k((I_k\setminus J_k)\setminus E)$, and in view of the definition of S_k this is at most

$$\mu_k((I_k\setminus J_k)\setminus E)\le\int_{(I_k\setminus J_k)\setminus E}\frac{S_k}{\sqrt{t}}dt\le S_k\sqrt{2^k}\Theta(2^{-k+1})\le S_k\theta_k2^{-k/2}.$$

Finally,

$$\mu_k(J_{k+1}\setminus E)\le\int_{J_{k+1}\setminus E}\frac{S_k}{\sqrt{t}}dt\le S_k\sqrt{2^{k+1}}\Theta(2^{-k})\le S_k\sqrt{2}\theta_k2^{-k/2}.$$

All these give

$$m_{k+1}\le\frac{1}{4\sqrt{2}}m_k+S_k\theta_k2^{-k/2}(1+\sqrt{2}). \tag{2.49}$$

Now (2.45) and (2.49) give a double recursive estimate for the pairs (S_k,m_k). Next we resolve this recursion. For $k\ge 2$ multiply (2.49) by

$$\begin{aligned}2B_1\theta_{k+1}2^{(k+1)/2} &= 2B_12^{k+1}\Theta((2+\alpha)2^{-k-1})2^{(k+1)/2}\\ &\le 4\sqrt{2}B_12^k\Theta((2+\alpha)2^{-k})2^{k/2}=4\sqrt{2}B_1\theta_k2^{k/2}\end{aligned}$$

to obtain

$$2B_1m_{k+1}\theta_{k+1}2^{(k+1)/2}\le B_1m_k\theta_k2^{k/2}+S_k\theta_k^24\sqrt{2}B_1(1+\sqrt{2}). \tag{2.50}$$

Therefore, if we define

$$L_k=S_k+2B_1m_k\theta_k2^{k/2},$$

then it follows from this last inequality and (2.45) that

$$\begin{aligned}L_{k+1} &\le S_k\left(1+B_1(1+4(2+\sqrt{2}))\theta_k^2\right)+2B_1m_k\theta_k2^{k/2}\\ &\le \left(1+B_1(1+4(2+\sqrt{2}))\theta_k^2\right)L_k.\end{aligned}$$

This is already in the form that can be iterated, and it follows from it that

$$L_k \le L_2 \prod_{j=2}^{k-1}\left(1 + B_1(1+4(2+\sqrt{2}))\theta_j^2)\right) \le L_2 \exp\left(B_2 \sum_{j=2}^{k-1} \theta_j^2\right),$$

where

$$B_2 = 5\cdot 10^4 \ge B_1(1+4(2+\sqrt{2})).$$

Finally, we get from (2.47), (2.46) and (2.50) that

$$\begin{aligned} L_2 &\le \frac{\sqrt{26}}{\pi}(1+B_1\theta_1^2) + \frac{4}{7} + 2B_1 m_2\theta_2 2 \\ &\le \frac{\sqrt{26}}{\pi}(1+B_1\theta_1^2) + \frac{4}{7} + B_1\left(\frac{\theta_1\sqrt{2}}{7} + \frac{\sqrt{26}}{\pi}\theta_1^2 4\sqrt{2}(1+\sqrt{2})\right) \\ &\le e^{B_2\theta_1^2}\left(\frac{\sqrt{26}}{\pi} + \frac{4}{7} + \frac{\sqrt{2}}{7}B_1\right) \le B_3 e^{B_2\theta_1^2} \end{aligned}$$

where $B_3 = 700$. This is what we get from the double recursion (2.45) and (2.49).

All these give

$$S_k \le L_k \le B_3 \exp\left(B_2 \sum_{j=1}^{k-1} \theta_j^2\right). \tag{2.51}$$

Suppose now that $S_j \le R_k$ for all $1 \le j \le k$. By the preceding inequality this is certainly true with

$$R_k = B_3 \exp\left(B_2 \sum_{j=1}^{k-1} \theta_j^2\right). \tag{2.52}$$

We obtain then from (2.49) and

$$\theta_j = \Theta((2+\alpha)2^{-j+1})2^j \le (2+\alpha)2^{-j+1}2^j = 4 + 2\alpha$$

that

$$m_{j+1} \le \frac{1}{4\sqrt{2}} m_j + R_k q 2^{-j/2} \quad \text{for } 1 \le j < k,$$

where $q = (4+2\alpha)(1+\sqrt{2})$. On iterating this we obtain from (2.46)

$$\begin{aligned} m_k &\le \left(\frac{1}{4\sqrt{2}}\right)^{k-1} m_1 + R_k q \left(\sum_{j=1}^{k-1} \frac{2^{-j/2}}{(4\sqrt{2})^{k-1-j}}\right) \\ &\le \left(\frac{1}{4\sqrt{2}}\right)^{k-1} \frac{1}{7} + \frac{R_k q}{2^{(k-1)/2}}\left(\sum_{l=0}^{\infty} \frac{1}{4^l}\right) \le \frac{1}{2^{(k-1)/2}}\left(\frac{4R_k q}{3} + \frac{1}{7}\right). \end{aligned} \tag{2.53}$$

With this we are ready to estimate the equilibrium measure μ_E. Let $0 < \delta \le 1/2$, and choose k so that $(1-\alpha)2^{-k} < \delta \le (1-\alpha)2^{-k+1}$. We have

$$\begin{aligned} \mu_E([0,\delta]) &\le \mu_E\left([0,(1-\alpha)2^{-k+1}]\right) \le \mu_{E\cup[0,(1-\alpha)2^{-k+1}]}([0,(1-\alpha)2^{-k+1}]) \\ &= \mu_k\left([0,(1-\alpha)2^{-k+1}]\right) + \mu_k\big|_{\widetilde{J_k}\setminus E}\left([0,(1-\alpha)2^{-k+1}]\right), \end{aligned}$$

where $\widetilde{\nu}$ denotes the balayage of a measure ν onto $E \cup [0,(1-\alpha)2^{-k+1}]$. Now it follows from this, from the choice of S_k, m_k and from (2.51)–(2.53) that

$$\begin{aligned} \mu_E([0,\delta]) &\leq R_k \int_0^{(1-\alpha)2^{-k+1}} \frac{dt}{\sqrt{t}} + m_k \qquad (2.54)\\ &\leq R_k \frac{2}{\sqrt{2^{k-1}}} + \frac{1}{2^{(k-1)/2}}\left(\frac{4R_k q}{3} + \frac{1}{7}\right) \leq B_4 \exp\left(B_2 \sum_{j=1}^{k-1} \theta_j^2\right)\sqrt{\delta}, \end{aligned}$$

where $B_2 = 5 \cdot 10^4$ and $B_4 = 2 \cdot 10^4$, and where we used that

$$\frac{1}{\sqrt{2^{k-1}}} \leq \sqrt{\frac{2}{1-\alpha}}\sqrt{\delta} \leq \sqrt{\frac{52}{25}}\sqrt{\delta}.$$

Finally, we can estimate the sum in the exponential function as

$$\begin{aligned} \sum_{j=1}^{k-1} \theta_j^2 &= 4\Theta^2(\frac{38}{39}) + \sum_{j=2}^{k-1} \Theta^2((2+\alpha)2^{-j})2^{2j}\\ &= 4\Theta^2(\frac{38}{39})\left(\int_{38/39}^{1} \frac{du}{u^3}\right) 2\frac{38^2}{77}\\ &+ \Theta^2\left(\frac{2+\alpha}{4}\right)\left(\int_{(2+\alpha)/4}^{38/39} \frac{du}{u^3}\right) 8\left[\left(\frac{4}{2+\alpha}\right)^2 - \left(\frac{39}{38}\right)^2\right]^{-1}\\ &+ \sum_{j=3}^{k-1} \Theta^2((2+\alpha)2^{-j})\left(\int_{(2+\alpha)2^{-j}}^{(2+\alpha)2^{-j+1}} \frac{1}{u^3} du\right) \frac{8(2+\alpha)^2}{3}\\ &\leq 40 \int_{(2+\alpha)2^{-k+1}}^{1} \frac{\Theta^2(u)}{u^3} du. \end{aligned}$$

Now this and (2.54) imply (2.12) for $r = \delta$ with $D = 40B_2 = 2 \cdot 10^6$, since $\delta \leq (2+\alpha)2^{-k+1}$.

■

2.4 Proof of Theorem 2.2

It will be convenient to prove first (2.8).

Set

$$E_1 = E^c = \{r \,|\, C_r(0) \cap E \neq \emptyset\} \subseteq [0,1].$$

Then $\Theta_E(t) = |[0,t] \setminus E_1|$ for all $t \in [0,1]$. For $R > 2$ and $|z| = R$ we have (c.f. (2.18))

$$\log(R-1) - \log \operatorname{cap}(E) \leq g_{\mathbf{C}\setminus E}(z) \leq \log(R+1) - \log \operatorname{cap}(E),$$

and a similar estimate is true for the set E_1. Thus, we obtain from the maximum principle that

$$g_{\mathbf{C}\setminus E}(0) \leq \log \frac{R+1}{\operatorname{cap}(E)} \omega(0, C_R(0), \Delta_R(0) \setminus E)$$

and

$$\omega(0, C_R(0), \Delta_R(0) \setminus E_1) \le \frac{1}{\log\big((R-1)/\mathrm{cap}(E)\big)} g_{\mathbf{C}\setminus E_1}(0).$$

Here, by Beurling's theorem [1, Theorem 3-6] , we have

$$\omega(0, C_R(0), \Delta_R(0) \setminus E) \le \omega(0, C_R(0), \Delta_R(0) \setminus E_1)$$

and so we obtain

$$g_{\mathbf{C}\setminus E}(0) \le \frac{\log\big((R+1)/\mathrm{cap}(E)\big)}{\log\big((R-1)/\mathrm{cap}(E)\big)} g_{\mathbf{C}\setminus E_1}(0),$$

and this yields for $R \to \infty$ that

$$g_{\mathbf{C}\setminus E}(0) \le g_{\mathbf{C}\setminus E_1}(0). \tag{2.55}$$

Thus, without loss of generality we may assume $E \subseteq [0,1]$, and then that E consists of a finite number of intervals.

First suppose that for the value $r \in [0, 1/2]$ we are considering we have the inequality

$$|[r, 2r] \cap E| \ge \frac{r}{2}. \tag{2.56}$$

Let

$$E_2 = (E \setminus [0, 2r]) \cup [3r/2, 2r].$$

It is known (see [27, Sec. IV.5.4]) that if

$$E = [a_1, b_1] \cup [a_2, b_2] \cup \cdots, \qquad a_i < b_i < a_{i+1},$$

is lying in $[0,1]$ and consists of a finite number of intervals, and we move to the right the leftmost interval $[a_1, b_1]$ into $[a_1', b_1']$ with $b_1' \le a_2$, and denote the so obtained set by E', i.e.

$$E' \subseteq [a_1', b_1'] \cup [a_2, b_2] \cup \cdots, \qquad b_1' - a_1' = b_1 - a_1,$$

then for $R \ge 1$

$$\omega(0, C_R(0), \Delta_R(0) \setminus E) \le \omega(0, C_R(0), \Delta_R(0) \setminus E').$$

On applying this finitely many times by moving always the leftmost subinterval of E lying in $[0, 2r]$ to the right into an interval lying in $[0, 2r]$ to close up gaps between consecutive subintervals of E lying in $[0, 2r]$, we obtain that

$$\omega(0, C_R(0), \Delta_R(0) \setminus E) \le \omega(0, C_R(0), \Delta_R(0) \setminus E_2)$$

since, by the assumption (2.56), we have $|[r, 2r] \cap E| \ge |[3r/2, 2r]|$. Making again comparison between the Green function and harmonic measures as before, we can conclude that $g_{\mathbf{C}\setminus E}(0) \le g_{\mathbf{C}\setminus E_2}(0)$. Furthermore, $\Theta_{E_2}(t) \le \Theta_E(t) + 3r/2$ for all $t \ge r$, and here the additional term $3r/2$ introduces only a constant on the right of (2.8), therefore (2.8) for E follows from (2.8) for E_2 (with a constant independent of r). Hence without loss of generality we may assume that

$$E \subseteq [3r/2, 1], \qquad \text{and} \qquad [3r/2, 2r] \subseteq E.$$

Let now

$$E_3 = E - \frac{3r}{2}.$$

Then

$$E_3 \subseteq [0,1], \qquad [0, r/2] \subseteq E_3$$

and

$$g_{\mathbf{C}\setminus E}(0) = g_{\mathbf{C}\setminus E_3}(-3r/2).$$

If μ_3 is the equilibrium measure of E_3, then the $E_3 \subseteq [0,1]$ special case of Theorem 2.3 (that has been proven above) yields for $t \in [0,1]$ the estimate

$$\mu_3([0,t]) \le D_1\sqrt{t} \tag{2.57}$$

with

$$D_1 = C_0 \exp\left(D_0 \int_{r/2}^1 \frac{\Theta_{E_3}^2(u)}{u^3} du\right) \tag{2.58}$$

where C_0 and D_0 are some absolute constants. In Theorem 2.3 the inner integral in (2.58) is actually on the interval $[t, 1]$, but since $\Theta_{E_3}(u) = 0$ for $u \in [0, r/2]$, the preceding estimate is correct regardless of t. This easily yields an estimate for the Green function (c.f. (2.18)):

$$\begin{aligned} g_{\mathbf{C}\setminus E_3}(-3r/2) &= g_{\mathbf{C}\setminus E_3}(-3r/2) - g_{\mathbf{C}\setminus E_3}(0) \\ &= \int \log \frac{3r/2 + t}{t} d\mu_3(t) = \int_0^{r/4} + \int_{r/4}^1 . \end{aligned}$$

For the second integral on the right we get

$$\int_{r/4}^1 \le \int_{r/4}^1 \frac{3r/2}{t} d\mu_3(t),$$

and integration by parts gives that this is at most

$$\frac{3r}{2}\frac{\mu_3([0,r/4])}{r/4} + \frac{3r}{2}\int_{r/4}^1 \frac{\mu_3([0,t])}{t^2} dt \le 9D_1\sqrt{r},$$

where, in the last step we used (2.57).

Again from (2.57) we get for the first integral that

$$\begin{aligned} \int_0^{r/4} &= \sum_{k=1}^\infty \int_{r/2^{k+2}}^{r/2^{k+1}} \log\frac{3r/2+t}{t} d\mu_3(t) \le \sum_{k=1}^\infty \log \frac{2r}{r/2^{k+2}} \mu_3([0, r/2^{k+1}]) \\ &\le D_1\sqrt{r}\sum_{k=1}^\infty \frac{k+3}{\sqrt{2^{k+1}}} \le 13 D_1\sqrt{r}. \end{aligned}$$

All these imply

$$g_{\mathbf{C}\setminus E}(0) \le 22 D_1 \sqrt{r}. \tag{2.59}$$

Since $\Theta_{E_3}(u) = \Theta_E(u + 3r/2)$ for all $u \in [0,1]$, we get

$$\begin{aligned} \int_{r/2}^1 \frac{\Theta_{E_3}^2(u)}{u^3} du &= \int_{r/2}^1 \frac{\Theta_E^2(u+3r/2)}{u^3} du = \int_{2r}^{1+3r/2} \frac{\Theta_E^2(u)}{(u-3r/2)^3} du \\ &\le \int_{2r}^{1+3r/2} \frac{\Theta_E^2(u)}{(u/4)^3} du \le 4^3\left(1 + \int_{2r}^1 \frac{\Theta_E^2(u)}{u^3} du\right), \end{aligned}$$

where in the last step we used that for $u \in [1, 1+3r/2]$ we have $\Theta_E(u) \le 2$. This shows that the number D_1 in (2.58) satisfies

$$D_1 \le C_0 e^{64} \exp\left(4^3 D_0 \int_{2r}^{1} \frac{\Theta_E^2(u)}{u^3} du\right),$$

and in view of (2.59) the estimate

$$g_{\mathbf{C}\setminus E}(0) \le C_2 \exp\left(D_2 \int_{2r}^{1} \frac{\Theta_E^2(u)}{u^3} du\right) \sqrt{r} \tag{2.60}$$

holds with some absolute constants C_2, D_2.

This derivation used the assumption (2.56). If (2.56) is not true, but its analogue with $2r$, i.e.

$$|[2r, 4r] \cap E| \ge \frac{1}{2} 2r \tag{2.61}$$

holds, then, as we have just proven,

$$g_{\mathbf{C}\setminus E}(0) \le C_2 \exp\left(D_2 \int_{4r}^{1} \frac{\Theta_E^2(u)}{u^3} du\right) \sqrt{2r} \tag{2.62}$$

is true. Furthermore, by the assumption that (2.56) is not true we have $\Theta_E(u) \ge r/2$ for all $u \in [2r, 4r]$, and so

$$\int_{2r}^{4r} \frac{\Theta_E^2(u)}{u^3} du \ge \int_{2r}^{4r} \frac{(r/2)^2}{u^3} du \ge \frac{3}{128}, \tag{2.63}$$

and since $\exp(D_2/32) > \sqrt{2}$ (we may assume D_2 to be bigger than, say 200), we get that the estimate in (2.62) is actually better than (2.60), what we want to prove.

In a similar manner, if (2.61) is still not true, but its analogue with $4r$, i.e.

$$|[4r, 8r] \cap E| \ge \frac{1}{2} 4r$$

is true, then on applying to this case the above proven result, we get again (2.60) just as before. Proceeding this way, we can complete the proof provided there is a k such that $2^k r \le 1/2$ and

$$[2^k r, 2^{k+1} r] \cap E| \ge \frac{1}{2} 2^k r. \tag{2.64}$$

Note that in this case the proof gives (2.8) without the factor $\log 1/\mathrm{cap}(E)$ on the right of (2.8).

If there is no such k, then (c.f. (2.63))

$$\sqrt{r} \exp\left(D_2 \int_{2r}^{1} \frac{\Theta_E^2(u)}{u^3} du\right) \ge \sqrt{r} \exp\left(D_2 \frac{1}{4} \frac{3}{128} \log \frac{1}{r}\right) \ge 1, \tag{2.65}$$

and in this case (2.8) follows from

$$g_{\mathbf{C}\setminus E}(0) = \log \frac{1}{\mathrm{cap}(E)} - U^{\mu_E}(0) \le \log \frac{2}{\mathrm{cap}(E)}$$

(here μ_E denotes the equilibrium measure of E).

This same estimate can be used for $1/2 < r \le 1$, as well, and the proof of (2.8) is complete.

Now we can complete the proof of Theorem 2.2. Since we can scale the set E, without loss of generality we may assume $E \subseteq \Delta_{1/2}(0)$, and $z = -r$, $0 < r < 1/2$. Let

$$E^c = \{r \,|\, C_r(0) \cap E \neq \emptyset\}$$

be the circular projection of E onto $[0,1]$. Beurling's theorem [1, Theorem 3-6] implies (see the proof of (2.55) above) that

$$g_{\mathbf{C}\setminus E}(-r) \le g_{\mathbf{C}\setminus E^c}(-r),$$

and so we may assume $E \subseteq [0,1]$. Now apply (2.8) for the set $E + r$ and for the origin. We obtain with some absolute constants C_0 and D_0

$$g_{\mathbf{C}\setminus E}(-r) \le C_0\sqrt{r}\exp\left(D_0\int_r^1 \frac{\Theta^2_{E,-r}(u)}{u^3}du\right)\log\frac{2}{\mathrm{cap}(E\cap[0,1-r])}.$$

Here

$$\Theta_{E,-r}(u)^2 \le (r + \Theta_E(u))^2 \le 2r^2 + 2\Theta^2_E(u),$$

and

$$\int_r^1 \frac{2r^2}{u^3}du \le 1.$$

Furthermore $E \cap [0, 1-r] = E$, and all these imply

$$g_{\mathbf{C}\setminus E}(-r) \le C_0 e^{D_0}\sqrt{r}\exp\left(2D_0\int_r^1 \frac{\Theta^2_E(u)}{u^3}du\right)\log\frac{2}{\mathrm{cap}(E)},$$

and this is (2.7).

The proof of (2.11) is similar if we recall that in the presence of a k with (2.64) there is no need for the factor $\log 1/\mathrm{cap}(E)$ in (2.2) (see the proof of Theorem 2.2). Now one can easily check that the condition (2.64) can be replaced for any fixed $\delta > 0$ by

$$[2^k r, 2^{k+1}r] \cap E| \ge \frac{\delta}{4}2^k r$$

for some k, and this latter one is easily implied by (2.10).

■

2.5 Proof of Theorem 2.1

Let E be a subset of the unit disk $\Delta_1(0)$, and let $E_1 = E \cap \Delta_{1/4}(0)$. For $|z| = 1$ we have

$$\log\frac{3}{4} + \log\frac{1}{\mathrm{cap}(E_1)} \le g_{\mathbf{C}\setminus E_1}(z)$$

which gives in view of $\log 3/(4\mathrm{cap}(E_1)) \ge \log 3 > 1$ that

$$g_{\mathbf{C}\setminus E_1}(z) \ge 1.$$

Thus, from the maximum principle we get for $|z| \leq 1$

$$\omega(z, C_1(0), \Delta_1(0) \setminus E) \leq \omega(z, C_1(0), \Delta_1(0) \setminus E_1) \leq g_{\mathbf{C}\setminus E_1}(z).$$

Now suppose that there is a k such that $2^k|z| \leq 1/4$ and

$$[2^k|z|, 2^{k+1}|z|] \cap E| \geq \frac{1}{2} 2^k|z|. \tag{2.66}$$

From the preceding proof we know that in this case the factor $\log 2/\text{cap}(E_1)$ can be dropped in the estimate (2.7) applied to the set E_1 rather than to E, i.e. in this case we obtain

$$\omega(z, C_1(0), \Delta_1(0) \setminus E) \leq C \exp\left(D \int_{|z|}^1 \frac{\Theta_{E_1}^2(u)}{u^3} du \right) \sqrt{|z|}.$$

Here $\Theta_{E_1}(u) = \Theta_E(u)$ for all $u \in [0, 1/4]$ and $\Theta_{E_1}(u) \leq 1$ for $u \in [1/4, 1]$ and so the preceding inequality implies a similar inequality with Θ_{E_1} replaced by Θ_E on the right, and that is exactly (2.3).

If there is no k with property (2.66), then (see (2.65))

$$\sqrt{|z|} \exp\left(D \int_{|z|}^1 \frac{\Theta_E^2(u)}{u^3} du \right) \geq \sqrt{|z|} \exp\left(D \frac{1}{4} \frac{1}{32} \log \frac{1}{|z|} \right) \geq 1,$$

and in this case (2.3) is trivially satisfied.

■

2.6 Proof of Theorem 2.3

Now we can complete the proof of Theorem 2.3 in the general case (recall that the case $E \subseteq [0, 1]$ has been verified in section 2.3). Clearly, if $C \geq 2^9$ and $r \geq 2^{-17}$, then (2.12) is true, so we may assume $r \leq 2^{-17}$.

Let $C_0 = C$ an $D_0 = D$ be the constants C, D from (2.11). We show that (2.12) is true with $D = 2D_0$ and $C = 40C_0 e^{2^{32} D_0}$.

In fact, suppose that $M > 40C_0 e^{2^{32} D_0}$ and that

$$\mu_E(\Delta_r(0)) < M\sqrt{r} \exp\left(2D_0 \int_r^1 \frac{\Theta_E^2(u)}{u^3} du \right) \tag{2.67}$$

is not true for some $0 < r \leq 1$. Let $r_0 \leq 1$ be the *largest* radius for which (2.67) is not true (this is certainly smaller than 1, actually much smaller than the bound 2^{-17} for r, because M is large), and for $r = r_0$ let us write the right hand side in (2.67) in the form $M\sqrt{r_0}L$ with

$$L = \exp\left(2D_0 \int_{r_0}^1 \frac{\Theta_E^2(u)}{u^3} du \right). \tag{2.68}$$

Thus,

$$\mu_E(\Delta_{r_0}(0)) \geq M\sqrt{r_0}L, \tag{2.69}$$

but for all $r_0 < r < 1$ we have

$$\mu_E(\Delta_r(0)) < M\sqrt{r} \exp\left(2D_0 \int_r^1 \frac{\Theta_E^2(u)}{u^3} du \right) \leq M\sqrt{r}L. \tag{2.70}$$

Since what we want to show is that r_0 does not exist, without loss of generality (by increasing M if necessary) we may assume that the equality sign holds in (2.69), i.e. that

$$\mu_E(\Delta_{r_0}(0)) = M\sqrt{r_0}L, \tag{2.71}$$

The lines $y = \pm x$ divide the plane into four quadrants Q_1, Q_2, Q_3 and Q_4, and for one of them, say for Q_j we must have

$$\mu_E(\Delta_{r_0}(0) \cap Q_j) \geq \frac{M}{4}\sqrt{r_0}L. \tag{2.72}$$

Without loss of generality let this happen for $j = 1$, and set (see Figure 2, where E is the union of the triangle and the quadrilateral, the small circle represents the disk $\Delta_{r_0}(0)$ and the large circle represents the disk $\Delta_{2^{16}r_0}(0)$)

$$E^* = (E \setminus \Delta_{2^{16}r_0}(0)) \bigcup (E \cap Q_1 \cap \Delta_{r_0}(0)).$$

We shall estimate the value $g_{\mathbf{C}\setminus E^*}(-r_0)$ of the Green function of $\mathbf{C} \setminus E^*$ in two ways, and these estimates will show that r_0 with the property (2.69)–(2.70) does not exist. This will show that (2.67) is indeed true.

First of all notice that with the equilibrium measure μ_{E^*} of E^*

$$g_{\mathbf{C}\setminus E^*}(-r_0) \geq g_{\mathbf{C}\setminus E^*}(-r_0) - g_{\mathbf{C}\setminus E^*}(0) = \int_{E^*} \log\left|\frac{w + r_0}{w}\right| d\mu_{E^*}(w).$$

Let us break the last integral into two parts: over the set $E^* \cap \Delta_{r_0}(0)$ and over the rest. First of all, for $w \in E^* \cap \Delta_{r_0}(0) = E \cap Q_1 \cap \Delta_{r_0}$ we have

$$\log\left|\frac{w + r_0}{w}\right| \geq \log\left|\frac{r_0 e^{i\pi/4} + r_0}{r_0 e^{i\pi/4}}\right| = \log(\sqrt{2 + \sqrt{2}}) > \frac{1}{2},$$

hence

$$I_2 := \int_{E^*\cap\Delta_{r_0}(0)} \log\left|\frac{w + r_0}{w}\right| d\mu_{E^*}(w) \geq \frac{1}{2}\mu_{E^*}(\Delta_{r_0}(0)).$$

Here

$$\mu_{E^*}(\Delta_{r_0}(0)) = \mu_{E^*}(\Delta_{r_0}(0) \cap Q_1) \geq \mu_E(\Delta_{r_0}(0) \cap Q_1)$$

because μ_{E^*} is the balayage of μ_E onto $E^* \subseteq E$, and $E^* \cap Q_1 \cap \Delta_{r_0}(0) = E \cap Q_1 \cap \Delta_{r_0}(0)$. Thus, from (2.72) we obtain that

$$I_2 \geq \frac{M}{10}\sqrt{r_0}L.$$

For $2^k r_0 \leq |w| \leq 2^{k+1}r_0$, $k \geq 16$ we have

$$\log\left|\frac{w + r_0}{w}\right| \geq \log\left|\frac{2^k r_0 - r_0}{2^k r_0}\right| = \log(1 - 2^{-k}) \geq -\frac{2}{2^k},$$

hence we obtain

$$\begin{aligned} I_1 &:= \int_{E^*\setminus\Delta_{r_0}(0)} \log\left|\frac{w + r_0}{w}\right| d\mu_{E^*}(w) = \sum_{k\geq 16} \int_{2^k r_0 \leq |w| \leq 2^{k+1}r_0} \\ &\geq \sum_{k\geq 16} \left(-\frac{2}{2^k}\right) \mu_{E^*}(\Delta_{2^{k+1}r_0}(0)). \end{aligned}$$

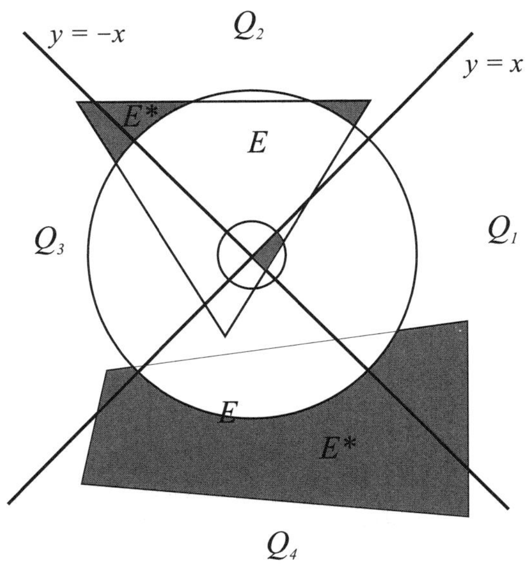

Figure 2:

Here (note that μ_{E^*} is the balayage of μ_E onto E^* and $E \setminus E^* \subseteq \Delta_{2^{16}r_0}(0)$)

$$\mu_{E^*}(\Delta_{2^{k+1}r_0}(0)) \le \mu_E(\Delta_{2^{k+1}r_0}(0)) + \mu_E(\Delta_{2^{16}r_0}(0)),$$

and so (2.70) yields

$$\mu_{E^*}(\Delta_{2^{k+1}r_0}(0)) \le M\sqrt{2^{k+1}r_0}L + M\sqrt{2^{16}r_0}L.$$

Thus, from the previous estimate it follows that

$$I_1 \ge -2M\sqrt{r_0}L \sum_{k=16}^{\infty} \left(\frac{\sqrt{2^{k+1}}}{2^k} + \frac{\sqrt{2^{16}}}{2^k} \right) \ge -\frac{M}{20}\sqrt{r_0}L.$$

All these imply

$$g_{\mathbf{C}\setminus E^*}(-r_0) = I_1 + I_2 \ge \frac{M}{10}\sqrt{r_0}L - \frac{M}{20}\sqrt{r_0}L = \frac{M}{20}\sqrt{r_0}L. \tag{2.73}$$

Next we invoke (2.7) for E^* and the point $z = -r_0$. Note that (2.71) implies that we must have a $k \ge 1$ with property

$$|[2^k r_0, 2^{k+1} r_0] \cap E^c| \ge \frac{1}{2} 2^k r_0 \tag{2.74}$$

(the analogue of (2.66)) where E^c is the circular projection of E onto $[0, 1]$. Indeed, otherwise the right hand side in (2.71) is bigger than 1 (c.f. (2.65) in the proof of Theorem 2.2). By the same token we may assume that there is a $k \ge 16$ for which (2.74) is true, and then for this same k we have

$$|[2^k r_0, 2^{k+1} r_0] \cap (E^*)^c| \ge \frac{1}{2} 2^k r_0.$$

Thus, by the remark made after Theorem 2.2, the inequality (2.11) (with C and D replaced by C_0 and D_0) holds and we obtain from (2.11)

$$g_{\mathbf{C}\setminus E^*}(-r_0) \leq C_0\sqrt{r_0}\exp\left(D_0\int_{r_0}^{1}\frac{\Theta_{E^*}^2(u)}{u^3}du\right).$$

On the right hand side here we have

$$\Theta_{E^*}^2(u) \leq \left(\Theta_E(u) + 2^{16}r_0\right)^2 \leq 2\Theta_E^2(u) + 2(2^{16}r_0)^2,$$

from which

$$D_0\int_{r_0}^{1}\frac{\Theta_{E^*}^2(u)}{u^3}du \leq 2D_0\int_{r_0}^{1}\frac{\Theta_E^2(u)}{u^3}du + 2D_0(2^{16}r_0)^2\int_{r_0}^{1}\frac{du}{u^3},$$

where the second term on the right is smaller than $2^{32}D_0$. Thus,

$$g_{\mathbf{C}\setminus E^*}(-r_0) \leq C_0e^{2^{32}D_0}\sqrt{r_0}\exp\left(2D_0\int_{r_0}^{1}\frac{\theta_E^2(u)}{u^3}du\right) = C_0e^{2^{32}D_0}\sqrt{r_0}L.$$

Now this and (2.73) cannot simultaneously hold since $M > 40C_0e^{2^{32}D_0}$, and this contradiction proves that, indeed, (2.67) is always true.

■

3 Sharpness

In this chapter we show that the theorems in Chapter 2 are best possible. One cannot expect a full converse in the sense that e.g. if

$$\int_0^1 \frac{\Theta_E^2(u)}{u^3} du = \infty \tag{3.1}$$

then (2.9) is not true. Indeed, disregarding the fact that $g_{\mathbf{C}\setminus E}(z)$ may be identically zero in a neighborhood of the origin, let us note also that there are compact subsets $E \subset [0,1]$ with zero linear measure for which the Green function $g_{\mathbf{C}\setminus E}$ is in Lip 1/2 (see e.g. Corollary 5.2 below), and for these E we have $\Theta_E(t) \equiv t$, i.e. (3.1) is true. For such E's, though (3.1) is true, E is dense at the origin in the sense that the complement of the circular projection of E contains in $[0,t]$ only very short intervals compared to t as $t \to 0$. However, regarding the structure of the estimates the theorems are sharp.

Theorem 3.1 *There are absolute constants $c, d > 0$ for which the following is true. If Θ is a nonnegative increasing function on $[0,1]$ with the property $\Theta(t) \le t$ for all t, then there is a compact set $E \subseteq [0,1]$ of positive capacity such that*

$$\Theta_E(t) \le \Theta(t) \qquad \text{for all} \quad t \in [0,1] \tag{3.2}$$

and for all $0 < r \le 1$ we have

$$\mu_E([0,r]) \ge c\sqrt{r} \exp\left(d \int_r^1 \frac{\Theta^2(u)}{u^3} du \right), \tag{3.3}$$

$$\omega(-r, C_1(0), \Delta_1(0) \setminus E) \ge c\sqrt{r} \exp\left(d \int_r^1 \frac{\Theta^2(u)}{u^3} du \right), \tag{3.4}$$

and

$$g_{\mathbf{C}\setminus E}(-r) \ge c\sqrt{r} \exp\left(d \int_r^1 \frac{\Theta^2(u)}{u^3} du \right). \tag{3.5}$$

The proof also yields the following, which gives a lower bound under fairly general conditions.

Corollary 3.2 *Let $\Theta(t) \le t$ be an increasing function on $[0,1]$, and let $1 = t_0 > t_1 > \cdots$ be a positive sequence such that*

$$0 < \alpha \le \frac{t_{k+1}}{t_k} \le \beta < 1 \tag{3.6}$$

are satisfied for all k with some $0 < \alpha < \beta < 1$. Suppose that $E \subseteq [0,1]$ is a set of positive capacity such that for some $\gamma > 0$ every interval $[t_k, t_{k-1}]$, $k \ge 1$ contains a subinterval of length $\ge \gamma(\Theta(t_k) - \Theta(t_{k+1}))$ that is disjoint from E. Then (3.4)–(3.5) hold for all $0 < r \le 1$ with some constants c, d.

If, in addition, for all $k \ge 1$ we have $|[t_k, t_{k-1}] \cap E| \ge \gamma t_k$, then (3.3) is also true.

Of course, one needs some additional assumption to achieve (3.3), for it may happen that E is very thin at 0, e.g. if $0 \notin E$ then $\mu_E([0,r]) = 0$ for small r.

The next corollary gives a necessary and sufficient condition for the Lip 1/2 property of Green functions for a wide class of compact subsets of the real line.

Corollary 3.3 *Let $1 = t_0 > t_1 > \cdots$ be a positive sequence with property (3.6), and let $E \subset [0,1]$ be a compact set. If $|[t_k, t_{k-1}] \setminus E| = \gamma_k t_k$ and*

$$\sum_{k=1}^{\infty} \gamma_k^2 < \infty, \tag{3.7}$$

then the Green function $g_{\mathbf{C}\setminus E}$ of $\mathbf{C} \setminus E$ satisfies a Lip 1/2 *condition at the origin, i.e. we have (2.9).*

Conversely, if for every k the set $[t_k, t_{k-1}] \setminus E$ contains an interval of length $\geq \gamma_k^* t_k$ with

$$\sum_{k=1}^{\infty} (\gamma_k^*)^2 = \infty, \tag{3.8}$$

then

$$\frac{g_{\mathbf{C}\setminus E}(-r)}{\sqrt{r}} \to \infty \tag{3.9}$$

as $r \to 0+0$.

Similar results hold for the harmonic measure $\omega(z, C_1(0), \Delta_1(0) \setminus E)$ and for the equilibrium measure μ_E. For example, assume that for all but finitely many k we have $|[t_k, t_{k-1}] \setminus E| \leq \gamma_k t_k$ with some $\gamma_j \leq \gamma < 1$ and that the set $[t_k, t_{k-1}] \setminus E$ contains an interval of length $\geq \gamma_k^* t_k$ with $\gamma_k^* \sim \gamma_k$. Then $\mu_E([0,\delta]) = O(\sqrt{\delta})$ is satisfied if and only if $\sum_k \gamma_k^2 < \infty$.

3.1 Proof of Theorem 3.1 and its corollaries

Let $\Theta(t) \leq t$ be an increasing nonnegative function on $[0,1]$. We set

$$E = \bigcup_{k=1}^{\infty} \left[\frac{1}{2^k} + \frac{1}{4}\left(\Theta\left(\frac{1}{2^k}\right) - \Theta\left(\frac{1}{2^{k+1}}\right)\right), \frac{2}{2^k}\right] \tag{3.10}$$

Since $\Theta(2^{-k})/4 \leq 2^{-k}/4$, the k-th interval in the definition of E contains at least the right 3/4 of the interval $[2^{-k}, 2^{-k+1}]$.

Let $t \leq 1$, $2^{-k} < t \leq 2^{-k+1}$. We have

$$\Theta_E(t) \leq \sum_{l=k}^{\infty} \frac{1}{4}\left(\Theta\left(\frac{1}{2^l}\right) - \Theta\left(\frac{1}{2^{l+1}}\right)\right) = \frac{1}{4}\Theta\left(\frac{1}{2^k}\right) \leq \Theta(t),$$

i.e. (3.2) is true for this E.

Let $d\mu_0(t) = dt/\pi\sqrt{t(1-t)}$ be the equilibrium measure of $[0,1]$, and for $k \geq 1$ let μ_k be the equilibrium measure of the set $E \cup [0, 2^{-k}]$. Then μ_k is absolutely continuous on $(0, 2^{-k})$ with respect to Lebesgue measure, and let S_k be the largest constant for which

$$\frac{d\mu_k(t)}{dt} \geq \frac{S_k}{\sqrt{t}} \quad \text{for } t \in \left(0, \frac{3}{4}2^{-k}\right]. \tag{3.11}$$

We set

$$\tau_k = \frac{1}{4}\left(\Theta(2^{-k}) - \Theta(2^{-k-1})\right), \qquad a_k = 2^{-k} + \frac{\tau_k}{2}.$$

Since $(2^{-k}, 2^{-k} + \tau_k) \cap E = \emptyset$, it follows that if we take the balayage of δ_{a_k} onto $E \cup [0, 2^{-k}]$, then on the interval $[0, 2^{-k}]$ this balayage measure is at least as large as the balayage measure $\widehat{\delta_{a_k}}$ that we obtain by taking the balayage of δ_{a_k} onto $[0, 2^{-k}] \cup [2^{-k} + \tau_k, 1]$. By (2.25) this latter balayage measure is given by the density

$$\frac{d\widehat{\delta_{a_k}}(t)}{dt} = \frac{1}{\pi}\frac{\sqrt{|a_k(a_k - 2^{-k})(a_k - (2^{-k} + \tau_k))(a_k - 1)|}}{\sqrt{|t(t - 2^{-k})(t - (2^{-k} + \tau_k))(t - 1)|}}\frac{|t - \zeta_k|}{|a_k - \zeta_k|}\frac{1}{|t - a_k|}, \tag{3.12}$$

where $\zeta_k \in (-\infty, 0) \cup (1, \infty)$ is the unique solution of the equation

$$\int_{-\infty}^{0} + \int_{1}^{\infty} H(t)(t - \zeta_k)dt = 0$$

with

$$H(t) = \frac{1}{\sqrt{|t(t - 2^{-k})(t - (2^{-k} + \tau_k))(t - 1)|}}\frac{1}{|t - a_k|}.$$

It is easy to see that ("$\sim$" meaning that the ratio of the two sides lies in between two constants)

$$\int_{1}^{\infty} H(t)dt \sim 1, \qquad \int_{-\infty}^{0} H(t)dt \sim 2^{3k/2},$$

$$\int_{1}^{\infty} tH(t)dt \sim 1, \qquad \int_{-\infty}^{0} tH(t)dt \sim -2^{k/2},$$

which imply $\zeta_k \sim -2^{-k}$, i.e.

$$-c_1 2^{-k} \le \zeta_k \le -c_2 2^{-k}$$

is true with some positive constants c_1, c_2. But then for $t \in (0, 3\cdot 2^{-k}/4]$ the density in (3.12) is

$$\sim \frac{\sqrt{2^{-k}\tau_k\tau_k 1}}{\sqrt{t 2^{-k} 2^{-k} 1}}\frac{2^{-k}}{2^{-k}}\frac{1}{2^{-k}} \sim \frac{\tau_k}{\sqrt{t}}(2^k)^{3/2},$$

i.e. there is a constant $c_3 > 0$ such that

$$\frac{d\widehat{\delta_{a_k}}(t)}{dt} \ge c_3\frac{\tau_k}{\sqrt{t}}(2^k)^{3/2}, \qquad t \in \left(0, \frac{3}{4}2^{-k}\right].$$

Since $d\widehat{\delta_a}(t)/dt$ is a harmonic function of a in the disk $|a - a_k| < \tau_k/2$, we obtain from the preceding inequality and from Harnack's inequality (2.30) that

$$\frac{d\widehat{\delta_a}(t)}{dt} \ge \frac{c_3}{3}\frac{\tau_k}{\sqrt{t}}(2^k)^{3/2} \qquad t \in \left(0, \frac{3}{4}2^{-k}\right] \tag{3.13}$$

for all $a \in [2^{-k} + \tau_k/4, 2^{-k} + 3\tau_k/4]$.

The equilibrium measure μ_k of $E \cup [0, 2^{-k}]$ is the balayage of μ_0 onto $E \cup [0, 2^{-k}]$. Forming this balayage measure can be done in two steps: first form the balayage of μ_0 onto $E \cup [0, 2^{-k+1}]$ to obtain μ_{k-1}, and then form the balayage of this μ_{k-1} onto $E \cup [0, 2^{-k}]$. In this last step we only have to form the balayage of $\mu_{k-1}\big|_{[2^{-k}, 2^{-k+1}] \setminus E} = \mu_{k-1}\big|_{(2^{-k}, 2^{-k} + \tau_k)}$ onto $E \cup [0, 2^{-k}]$ and add this balayage measure to $\mu_{k-1}\big|_{E \cup [0, 2^{-k}]}$. Thus, by (3.13) we have for $t \in (0, 3 \cdot 2^{-k}/4]$ the estimate

$$\frac{d\mu_k(t)}{dt} \geq \frac{d\mu_{k-1}(t)}{dt} + \int_{2^{-k}+\tau_k/4}^{2^{-k}+3\tau_k/4} \frac{c_3}{3} \frac{\tau_k}{\sqrt{t}} (2^k)^{3/2} d\mu_{k-1}(a).$$

Since $\tau_k \leq 2^{-k}/4$, it follows from the definition of the number S_{k-1} that we have for

$$t \in (0, 3 \cdot 2^{-k}/4] \qquad \text{and} \qquad a \in [2^{-k} + \tau_k/4, 2^{-k} + 3 \cdot \tau_k/4] \subseteq (0, 3 \cdot 2^{-k+1}/4]$$

the estimates

$$\frac{d\mu_{k-1}(t)}{dt} \geq \frac{S_{k-1}}{\sqrt{t}}, \qquad \frac{d\mu_{k-1}(a)}{da} \geq \frac{S_{k-1}}{\sqrt{a}},$$

hence it follows that

$$\begin{aligned} \frac{d\mu_k(t)}{dt} &\geq \frac{S_{k-1}}{\sqrt{t}} + \frac{c_3}{3} \frac{\tau_k}{\sqrt{t}} (2^k)^{3/2} \frac{\tau_k}{2} \frac{S_{k-1}}{\sqrt{2 \cdot 2^{-k}}} \\ &= \frac{S_{k-1}}{\sqrt{t}} \left(1 + \frac{c_3}{6\sqrt{2}} \tau_k^2 2^{2k}\right), \end{aligned}$$

i.e.

$$S_k \geq S_{k-1} \left(1 + \frac{c_3}{9} \tau_k^2 2^{2k}\right).$$

This can be iterated and we obtain

$$S_k \geq \prod_{l=2}^{k} \left(1 + \frac{c_3}{9} \tau_l^2 2^{2l}\right) S_1 \geq \frac{1}{\pi} \exp\left(\frac{c_3}{18} \sum_{l=2}^{k} \tau_l^2 2^{2l}\right).$$

Here with $s_l = 2^l \Theta(2^{-l}) \leq 1$ we have

$$\begin{aligned} 4^2 \sum_{l=2}^{k} \tau_l^2 2^{2l} &= \sum_{l=2}^{k} \left(s_l - \frac{1}{2} s_{l+1}\right)^2 \\ &= \sum_{l=2}^{k} s_l^2 - \sum_{l=2}^{k-1} s_l s_{l+1} - s_k s_{k+1} + \frac{1}{4} \sum_{l=2}^{k} s_{l+1}^2 \geq \frac{1}{4} \sum_{l=2}^{k} s_{l+1}^2 - 1, \end{aligned}$$

where in the last step we used that, by the Cauchy-Schwarz inequality,

$$\sum_{l=2}^{k} s_l^2 - \sum_{l=2}^{k-1} s_l s_{l+1} \geq 0.$$

Furthermore,

$$\begin{aligned} \sum_{l=2}^{k} s_{l+1}^2 &\geq \sum_{l=3}^{k} \Theta(2^{-l})^2 2^{2l} \geq \frac{1}{2} \int_{2^{-k-1}}^{1/8} \frac{\Theta^2(u)}{u^3} du \\ &\geq \frac{1}{2} \int_{2^{-k-1}}^{1} \frac{\Theta^2(u)}{u^3} du - 16. \end{aligned}$$

All these imply

$$S_k \geq \frac{1}{\pi} e^{-1} \exp\left(\frac{c_3}{18 \cdot 128} \int_{2^{-k-1}}^{1} \frac{\Theta^2(u)}{u^3} du \right). \tag{3.14}$$

Now we can prove that (3.3) holds for this E. Let $0 < r \leq 1$, $2^{-k} < r \leq 2^{-k+1}$. Since $[5 \cdot 2^{-k}/8, 3 \cdot 2^{-k}/4] \subset E$ and μ_E is the balayage of μ_k onto E, we get

$$\mu_E\left(\left[\frac{5}{8}2^{-k}, \frac{3}{4}2^{-k}\right]\right) \geq \mu_k\left(\left[\frac{5}{8}2^{-k}, \frac{3}{4}2^{-k}\right]\right).$$

Thus, the just obtained estimate on S_k yields

$$\begin{aligned} \mu_E([0,r]) &\geq \mu_E\left(\left[\frac{5}{8}2^{-k}, \frac{3}{4}2^{-k}\right]\right) \geq \mu_k\left(\left[\frac{5}{8}2^{-k}, \frac{3}{4}2^{-k}\right]\right) \geq \int_{5 \cdot 2^{-k}/8}^{3 \cdot 2^{-k}/4} \frac{S_k}{\sqrt{t}} dt \\ &\geq \frac{S_k}{\sqrt{2^{-k}}} \frac{2^{-k}}{8} \geq \frac{\sqrt{r}}{100} \exp\left(\frac{c_3}{18 \cdot 128} \int_r^1 \frac{\Theta^2(u)}{u^3} du \right), \end{aligned}$$

where we have used (3.14) and that $2^{-k-1} \leq r$ and $\sqrt{2^{-k}} \geq \sqrt{r}/\sqrt{2}$. This proves (3.3).

(3.5) is an immediate consequence of (3.3):

$$\begin{aligned} g_{\mathbf{C}\setminus E}(-r) &= g_{\mathbf{C}\setminus E}(-r) - g_{\mathbf{C}\setminus E}(0) = \int \log \frac{t+r}{t} d\mu_E(t) \\ &\geq \int_{[0,r]} \log 2 \, d\mu_E(t) \geq (\log 2) c \sqrt{r} \exp\left(d \int_r^1 \frac{\Theta^2(u)}{u^3} du \right). \end{aligned}$$

Finally, (3.4) is a consequence of (3.5) and of the maximum principle, since both $\omega(z, C_1(0), \Delta_1(0) \setminus E)$ and $g_{\mathbf{C}\setminus E}(z)$ are harmonic functions in $D_1(0) \setminus E$, and for $|z| = 1$ we have

$$\omega(z, C_1(0), \Delta_1(0) \setminus E) = 1 \geq \frac{1}{4}\left(\log 2 + \log \frac{1}{\operatorname{cap}(E)} \right) \geq \frac{1}{4} g_{\mathbf{C}\setminus E}(z).$$

■

In proving Corollary 3.2 note first of all that if E^* is any set of the form

$$E^* = \bigcup_{k=1}^{\infty} [t_k, t_{k-1}] \setminus I_k, \tag{3.15}$$

where $I_k \subset [t_k, t_{k-1}]$ is an interval of length $\gamma(\Theta(t_k) - \Theta(t_{k+1}))/4$, then (3.4)–(3.5) are true with some constants c, d depending only on γ. In fact, little changes are needed in the preceding proof; in the present case $\{t_k\}$ plays the role of the sequence $\{2^{-k}\}$.

Now since both $\omega(z, C_1(0), \Delta_1(0) \setminus E)$ and $g_{\mathbf{C}\setminus E}(z)$ are decreasing functions of E, the statements regarding (3.4) and (3.5) immediately follow, since the E in Corollary 3.2 is contained in a set of the form (3.15). Finally, if $|E \cap [t_k, t_{k-1}]| \geq \gamma t_k$

is also satisfied with some $\gamma > 0$, then to prove (3.3) we follow the proof of (3.3) above, but complete it in the following way. The proof of (3.3) gives that for $t \in (0, (t_{k+1} + t_k)/2]$ the inequality

$$\frac{d\mu_k(t)}{dt} \geq \frac{S_k}{\sqrt{t}},$$

holds, where now μ_k is the equilibrium measure of the set $E \cup [0, t_k]$, and for S_k we have the estimate (c.f. (3.14))

$$S_k \geq c \exp\left(d \int_{t_k}^1 \frac{\Theta^2(u)}{u^3} du\right). \tag{3.16}$$

with some positive constants c and d.

Let $0 < r \leq 1$, $t_k < r \leq t_{k-1}$. Since

$$E \cap [t_{k+2}, t_{k+1}] \subset [0, (t_k + t_{k+1})/2],$$

we get

$$\begin{aligned}\mu_E([0,r]) &\geq \mu_E(E \cap [t_{k+2}, t_{k+1}]) \geq \mu_k(E \cap [t_{k+2}, t_{k+1}]) \geq \int_{E \cap [t_{k+2}, t_{k+1}]} \frac{S_k}{\sqrt{t}} dt \\ &\geq \frac{S_k}{\sqrt{t_{k+1}}} \gamma t_{k+2} \geq c\sqrt{r} \exp\left(d \int_r^1 \frac{\Theta^2(u)}{u^3} du\right),\end{aligned}$$

since $t_{k+1} \sim t_{k+2} \sim r$. This proves (3.3), and the proof of Corollary 3.2 is complete. ∎

Corollary 3.3 can be deduced from Theorem 2.2 and Corollary 3.2. In fact, note that under the assumptions of Corollary 3.3 we have

$$\Theta_E(t_k) = \sum_{j=k+1}^{\infty} \gamma_j t_j. \tag{3.17}$$

Suppose that $\sum_k \gamma_k^2 < \infty$. Then using (3.17), the fact that $t_j/t_k \leq \beta^{j-k}$ for all $j > k$ (see (3.6)) and the Cauchy-Schwarz inequality we see that

$$\begin{aligned}\int_0^1 \frac{\Theta(u)^2}{u^3} du &\sim \sum_k \Theta(t_k)^2 t_k^{-2} = \sum_k \left(\sum_{j=k+1}^{\infty} \gamma_j \frac{t_j}{t_k}\right)^2 \\ &\leq C \sum_k \sum_{j=k+1}^{\infty} \gamma_j^2 \left(\frac{t_j}{t_k}\right)^2 \leq C \sum_k \gamma_k^2 < \infty,\end{aligned}$$

and so the Lip 1/2 property of $g_{\mathbf{C}\setminus E}$ at the origin follows from Theorem 2.2 (c.f. (2.9)).

Conversely, let each $[t_k, t_{k+1}] \setminus E$ contain an interval of length $\geq \gamma_k^* t_k$. If we set

$$\Theta(t) = \sum_{j=k}^{\infty} \gamma_j^* t_j$$

for $t \in [t_k, t_{k-1})$, then for E the condition of Corollary 3.2, namely that $[t_k, t_{k-1}] \setminus E$ contains an interval of length $\geq \gamma(\Theta(t_k) - \Theta(t_{k+1}))$, is satisfied with $\gamma = 1$. Thus, if $\sum_k (\gamma_k^*)^2 = \infty$, which implies

$$\int_0^1 \frac{\Theta(u)^2}{u^3} du \sim \sum_k \Theta(t_k)^2 t_k^{-2} \geq \sum_k (\gamma_k^*)^2 = \infty,$$

then it follows from Corollary 3.2, (3.5) that for $g_{\mathbf{C}\setminus E}$ we have (3.9).

■

4 Higher order smoothness

Until now we have considered $\sqrt{|z|}$ order of smoothness for the Green functions and harmonic measures. In this chapter we show that the results from Chapter 2 can be used to detect higher order of smoothness. The claims here depend on the results from the previous section and on a symmetrization theorem of Baernstein [10].

First we mention a result that we shall use for sets lying on the real line and that uses less hypothesis than our main theorem. By simple symmetrization one can get from the results of Section 2.1 conditions that provide Lip 1 behavior of Green functions and harmonic measures. In fact, let $E \subseteq [-1,1]$, and let us consider its two-sided density

$$\Theta_E^*(t) = |[-t,t] \setminus E|. \tag{4.1}$$

Now the following corollary easily follows from the results in Chapter 2.

Corollary 4.1 *There is an absolute constants C such that if E is a compact subset of $[-1,1]$ of positive capacity and $0 < |z| < 1$, then*

$$\omega(z, C_1(0), \Delta_1(0) \setminus E) \leq C|z| \exp\left(C \int_{|z|}^1 \frac{\Theta_E^*(u)^2}{u^3} du\right), \tag{4.2}$$

$$g_{\mathbf{C}\setminus E}(z) \leq C|z| \exp\left(C \int_{|z|}^1 \frac{\Theta_E^*(u)^2}{u^3} du\right) \log \frac{2}{\mathrm{cap}(E)}, \tag{4.3}$$

and for $0 < r < 1$

$$\mu_E([-r,r]) \leq Cr \exp\left(C \int_r^1 \frac{\Theta_E^*(u)^2}{u^3} du\right). \tag{4.4}$$

In particular, if

$$\int_0^1 \frac{\Theta_E^*(u)^2}{u^3} du < \infty,$$

then we get a local Lip 1 behavior.

Corollary 4.1 is sharp in the sense of the preceding section. In fact, the following corollary easily follows from the results of Section 3.

Corollary 4.2 *There is an absolute constant $c > 0$ for which the following is true. If Θ^* is a nonnegative increasing function on $[0,1]$ with the property $\Theta^*(t) \leq 2t$ for all t, then there is a compact set $E \subseteq [-1,1]$ of positive capacity such that E is symmetric onto the origin,*

$$\Theta_E^*(t) \leq \Theta^*(t) \qquad \text{for all} \quad t \in [0,1] \tag{4.5}$$

and for all $0 < r \leq 1$ we have

$$\mu_E([-r,r]) \geq cr \exp\left(c \int_r^1 \frac{\Theta^*(u)^2}{u^3} du\right), \tag{4.6}$$

$$\omega(-r, C_1(0), \Delta_1(0) \setminus E) \geq cr \exp\left(c \int_r^1 \frac{\Theta^*(u)^2}{u^3} du\right), \tag{4.7}$$

and

$$g_{\mathbf{C}\setminus E}(-r) \geq cr \exp\left(c \int_r^1 \frac{\Theta^*(u)^2}{u^3} du\right). \tag{4.8}$$

After that let again $E \subset \mathbf{C}$ be a compact set on the complex plane, and for $0 < \alpha < 1$ consider the set

$$E_\alpha = \{r \mid |E \cap C_r(0)| \geq 2r\pi\alpha\} \tag{4.9}$$

of those r for which the circle $C_r(0)$ about 0 and of radius r intersects E in a set of (arc) measure $\geq 2\pi r\alpha$. The $\alpha = 0$ case would correspond to the circular projection of E onto $[0, \infty)$.

Theorem 4.3 *There is an absolute constant C depending only on α such that if E is a compact subset of the unit disk of positive capacity and $0 < |z| < 1$, then*

$$\omega(z, C_1(0), \Delta_1(0) \setminus E) \leq C|z|^{1/2(1-\alpha)} \exp\left(C \int_{|z|}^1 \frac{\Theta_{E_\alpha}^2(u)}{u^3} du\right), \tag{4.10}$$

$$g_{\mathbf{C}\setminus E}(z) \leq C|z|^{1/2(1-\alpha)} \exp\left(C \int_{|z|}^1 \frac{\Theta_{E_\alpha}^2(u)}{u^3} du\right) \log \frac{2}{\operatorname{cap}(E)}, \tag{4.11}$$

and for $0 < r < 1$

$$\mu_E(\Delta_r(0)) \leq Cr^{1/2(1-\alpha)} \exp\left(C \int_r^1 \frac{\Theta_{E_\alpha}^2(u)}{u^3} du\right). \tag{4.12}$$

In particular, if

$$\int_0^1 \frac{\Theta_{E_\alpha}(t)^2}{t^3} dt < \infty, \tag{4.13}$$

then both $\omega(z, C_1(0), \Delta_1(0) \setminus E)$ and $g_{\mathbf{C}\setminus E}(z)$ satisfy a local Lip $1/2(1-\alpha)$ smoothness at the origin.

Proof of Corollary 4.1. Let $E^* = E \cap (-E)$. Then E^* is symmetric onto the origin and

$$\Theta_{E^*}(t) \leq \Theta_E^*(t) \leq 2\Theta_{E^*}(t) \tag{4.14}$$

for all t, where Θ_{E^*} is the function (2.1) for the set E^* and the point $P = 0$. Let us apply the mapping $T(z) = z^2$, and let F be the image of E^* under this mapping. Since

$$\omega(z, C_1(0), \Delta_1(0) \setminus E) \leq \omega(z, C_1(0), \Delta_1(0) \setminus E^*) = \omega(z^2, C_1(0), \Delta_1(0) \setminus F),$$

we obtain from (2.2)

$$\omega(z, C_1(0), \Delta_1(0) \setminus E) \leq C\sqrt{|z^2|} \exp\left(C \int_{|z^2|}^1 \frac{\Theta_F^2(u)}{u^3} du\right).$$

Since F is the image set of E^* under the mapping $t \to t^2$, we can apply Lemma 11.5 from the Appendix to conclude (4.2) (recall also (4.14) that $\Theta_{E^*}(t)$ and $\Theta_E^*(t)$ have the same order).

In a similar manner follows (4.3) from (2.7).

Finally, it is easy to show (again with the mapping $x \to x^2$) that for the equilibrium measures we have $\mu_F([0, r^2]) = \mu_{E^*}([-r, r])$, and so

$$\mu_{E^*}([-r, r]) \leq Cr \exp\left(C \int_r^1 \frac{\Theta_{E^*(u)}^2}{u^3} du\right) \tag{4.15}$$

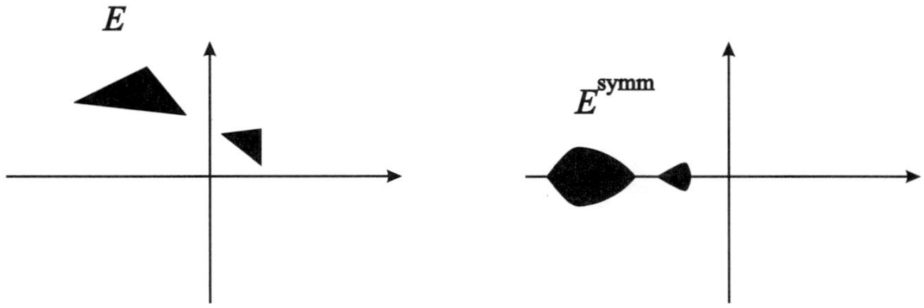

Figure 3:

follows from (2.12). Now let $E_r = E \cup [-r, r]$. Then $\mu_{(E_r)^*}$ is the balayage of μ_{E_r} onto $(E_r)^*$, hence

$$\mu_{E_r}([-r, r]) \leq \mu_{(E_r)^*}([-r, r])$$

because $[-r, r] \subset (E_r)^*$. On the other hand, μ_E is the balayage of μ_{E_r} onto E, hence

$$\mu_E([-r, r]) \leq \mu_{E_r}([-r, r])$$

because $E \setminus [-r, r] = E_r \setminus [-r, r]$. Now apply (4.15) with E replaced by E_r. The resulting inequality and the preceding two inequalities give

$$\mu_E([-r, r]) \leq Cr \exp\left(C \int_r^1 \frac{\Theta_{(E_r)^*}(u)^2}{u^3} du \right),$$

and since here $\Theta_{(E_r)^*}(u) \leq 2\Theta_E^*(u)$, the proof of (4.2) is complete.

■

Proof of Theorem 4.3. First of all, (4.12) follows from (4.11) exactly as we proved Theorem 2.3 from Theorem 2.2 in Subsection 2.6. Also, the inequality (4.10) follows from (4.11) (see the appropriate proofs in Section 2). Thus, it is enough to prove (4.11).

The proof uses Baernstein's [10] circular symmetrization result. If E is a compact subset of the unit disk, then its circular symmetrization set E^{symm} is defined as follows: for any $0 \leq r < 1$ let $\sigma(r)r$ denote the arc measure of the intersection $E \cap C_r(0)$. If this intersection is empty, then so is the intersection $E^{\mathrm{symm}} \cap C_r(0)$. Otherwise let

$$E^{\mathrm{symm}} \cap C_r(0) = C_r(0) \setminus \{re^{i\theta} \mid -\pi + \sigma(r)/2 < \theta < \pi - \sigma(r)/2\}$$

(see Figure 3). Thus, the intersection of E and of E^{symm} with any circle $C_r(0)$ have the same arc measures. Now a fundamental theorem of Baernstein [10, Theorem 7] says that for $|z| < 1$ we have

$$\omega(z, C_1(0), \Delta_1(0) \setminus E) \leq \omega(|z|, C_1(0), \Delta_1(0) \setminus E^{\mathrm{symm}}). \tag{4.16}$$

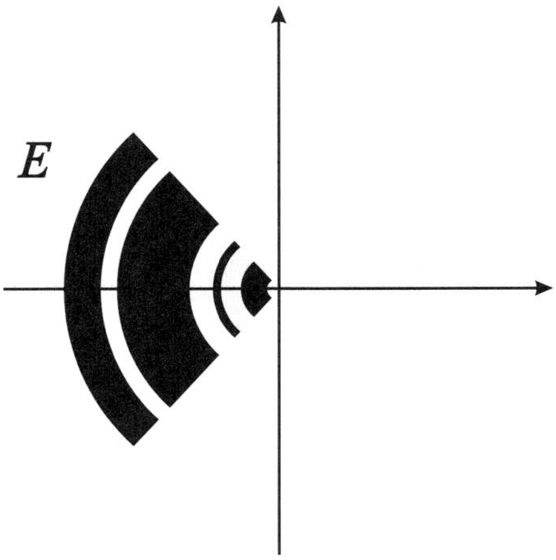

Figure 4:

This easily implies the same inequality for Green functions, i.e.

$$g_{\mathbf{C}\setminus E}(z) \leq g_{\mathbf{C}\setminus E^{\mathrm{symm}}}(|z|). \tag{4.17}$$

Clearly,

$$\Theta_{E_\alpha}(t) = \Theta_{(E^{\mathrm{symm}})_\alpha}(t),$$

therefore in what follows we may assume $E = E^{\mathrm{symm}}$. In Θ_{E_α} we count only those r for which $C_r(0)$ intersects E in a measure $\geq 2\pi r\alpha$, therefore we may assume that for each r either $E \cap C_r(0) = \emptyset$ or $E \cap C_r(0)$ has arc measure exactly $2\pi\alpha$ (note also that by making E smaller we increase $g_{\mathbf{C}\setminus E}(z)$). Thus, in what follows we shall assume that E consists of some circular arcs on the circles $C_r(0)$ running from the point $re^{\pi(1-\alpha)}$ to $re^{-i\pi(1-\alpha)}$ and crossing the negative real line (see Figure 4), and all we have to do is to estimate the values $g(r) = g_{\mathbf{C}\setminus E}(r)$, $0 < r < 1$. We shall do that by reducing the problem by a transformation to Theorem 2.1. Let us immediately mention that E is a regular set, i.e. the Green function $g_{\mathbf{C}\setminus E}(z)$ of $\mathbf{C} \setminus E$ is continuous on the whole plane. In fact, this is clear if $0 \notin E$, for then every point of E is contained in an arc lying in E. But if $0 \in E$ then 0 is also a regular boundary point of $\mathbf{C} \setminus E$ because Wiener's criterion (2.19) holds at 0 (note that the capacity of the arc running from $re^{\pi(1-\alpha)}$ to $re^{-i\pi(1-\alpha)}$ is $\geq cr$ with some $c > 0$ that depends only on α).

First let us assume that $0 < \alpha \leq 1/2$. In that case the transformation we use is $z \to z^{1/(1-\alpha)}$, where $z^{1/(1-\alpha)}$ is defined on the plane cut along the negative real line and where we take that branch of $z \to z^{1/(1-\alpha)}$ that is positive for positive z. This maps the sector

$$S_\alpha = \{re^{it} \mid -\pi(1-\alpha) \leq t \leq \pi(1-\alpha)\} \tag{4.18}$$

onto the complex plane with the negative real line covered twice. The image $\tilde{E}$ of the set $E \cap S_\alpha$ under this mapping is the set of those points $-r^{1/(1-\alpha)}$ for which $C_r(0) \cap E \neq \emptyset$ (see Figure 5). By the symmetry of the set E we have

$$g(re^{i\pi(1-\alpha)}) = g(re^{-i\pi(1-\alpha)}),$$

therefore the function $G(w) = g(w^{1-\alpha})$ is well defined on the complex plane and it is continuous there. Here $w^{1-\alpha} = |w|^{1-\alpha}e^{i\mathrm{Arg}(w)}$ with the main branch of the argumentum function: $\mathrm{Arg}(w) \in (-\pi, \pi]$.

We are going to show that $G(w)$ is a subharmonic function on the complex plane. This will already prove the theorem as follows. Since $G(w) = (1-\alpha)\log|w| + O(1)$ as $w \to \infty$ and $G(w)$ is bounded on compact subsets of $\mathbf{C}$, we can see that the function $G(w) - g_{\mathbf{C}\setminus\tilde{E}}(w)$ is subharmonic on $\mathbf{C}\setminus\tilde{E}$, tends to $-\infty$ as $w \to \infty$ and it is and bounded from above on $\overline{\mathbf{C}}\setminus\tilde{E}$. Furthermore, by the continuity of the Green function $g_{\mathbf{C}\setminus E}$ we have $G(w) = 0$ for $w \in \tilde{E}$, therefore for any $w_0 \in \tilde{E}$

$$\limsup_{w\to w_0,\ w\notin\tilde{E}} G(w) - g_{\mathbf{C}\setminus\tilde{E}}(w) \le 0.$$

These imply by the subharmonicity of $G(w) - g_{\mathbf{C}\setminus\tilde{E}}(w)$ in $\mathbf{C}\setminus\tilde{E}$ that this is a nonpositive function (use the maximum principle for subharmonic functions on a disk of large radius), i.e.

$$G(w) \le g_{\mathbf{C}\setminus\tilde{E}}(w).$$

Now apply to the Green function $g_{\mathbf{C}\setminus\tilde{E}}(w)$ Theorem 2.2 to conclude

$$G(w) \le C\sqrt{|w|}\exp\left(D\int_{|w|}^{1}\frac{\Theta_{\tilde{E}}^2(u)}{u^3}du\right).$$

Substituting here $w = |z|^{1/(1-\alpha)}$ we get for all z

$$g_{\mathbf{C}\setminus E}(z) \le g_{\mathbf{C}\setminus E}(|z|) \le C\sqrt{|z|^{1/(1-\alpha)}}\exp\left(D\int_{|z|^{1/(1-\alpha)}}^{1}\frac{\Theta_{\tilde{E}}^2(u)}{u^3}du\right).$$

Here the set $\tilde{E}$ is obtained from $E\cap(-\infty,0]$ by the transformation $x \to -|x|^{1/1-\alpha}$, and in the Appendix, in Lemma 11.5 we show how the Θ integral is transformed under such transformation. Thus, an application of Lemma 11.5 gives

$$g_{\mathbf{C}\setminus E}(z) \le C|z|^{1/2(1-\alpha)}\exp\left(D\int_{|z|}^{1}\frac{\Theta_{E}^2(u)}{u^3}du\right),$$

and this is what we needed to prove.

Thus, it has remained to show the subharmonicity of G, and since G is bounded from above in the unit disk, it is enough to show the subharmonicity of G on $\mathbf{C}\setminus\{0\}$. But G is harmonic off the negative real line, it is nonnegative and it vanishes on $\tilde{E}$, hence all we have to show is the subharmonicity of G at points of $(-\infty,0)\setminus\tilde{E}$, and to this end it is enough to show that if $w_0 \in (-\infty,0)\setminus\tilde{E}$, then for sufficiently small $\rho > 0$

$$G(w_0) \le \frac{1}{2\pi}\int_{-\pi}^{\pi} G(w_0+\rho e^{it})dt. \tag{4.19}$$

By the symmetry of the function G the last integral is the same as (see Figure 5)

$$\frac{1}{\pi}\int_{0}^{\pi} G(w_0+\rho e^{it})dt = \frac{1}{\pi}\int_{0}^{\pi} g\left((w_0+\rho e^{it})^{1-\alpha}\right)dt,$$

where we use the branch of $w^{1-\alpha}$ that is positive for positive w. If $U(w) = w^{1-\alpha}$ with this branch, then, since the (upper part of the) negative real line is mapped

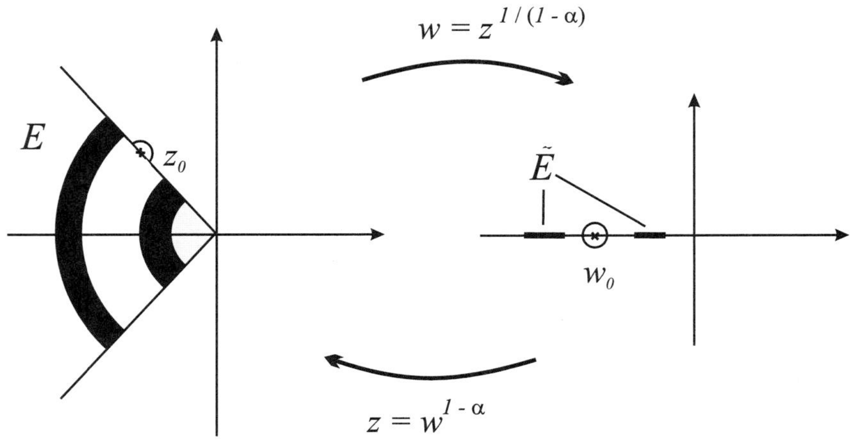

Figure 5:

by U into the ray $\{r^{i\pi(1-\alpha)} \mid 0 < r < \infty\}$, the argument of $U'(w_0)$ is $-\pi\alpha$. Thus, with $z_0 = w_0^{1-\alpha}$

$$\begin{aligned}(w_0 + \rho e^{it})^{1-\alpha} &= U(w_0 + \rho e^{it}) = U(w_0) + U'(w_0)\rho e^{it} + O(\rho^2) \\ &= z_0 + A\rho e^{i(t-\alpha\pi)} + O(\rho^2),\end{aligned}$$

where $A > 0$. Let now h be an analytic function in a neighborhood of z_0 with real part equal to $g(z) = g_{\mathbf{C}\setminus E}(z)$. Then $h'(z_0) = g_x(z_0) - ig_y(z_0)$, and hence

$$\begin{aligned}G(w_0 + \rho e^{it}) &= g(U(w_0 + \rho e^{it})) = \Re h(z_0 + A\rho e^{i(t-\alpha\pi)} + O(\rho^2)) \\ &= \Re\{h(z_0) + h'(z_0)A\rho e^{i(t-\alpha\pi)} + O(\rho^2)\} \\ &= g(z_0) + A\rho\,(g_x(z_0)\cos(t-\alpha\pi) + g_y(z_0)\sin(t-\alpha\pi)) + O(\rho^2).\end{aligned}$$

All these give

$$\frac{1}{2\pi}\int_{-\pi}^{\pi} G(w_0 + \rho e^{it})dt = G(w_0) + \frac{A\rho}{\pi}B + O(\rho^2), \tag{4.20}$$

where

$$B = g_x(z_0)(\sin(\pi - \alpha\pi) - \sin(-\alpha\pi)) + g_y(z_0)(\cos(-\alpha\pi) - \cos(\pi - \alpha\pi)).$$

Elementary trigonometry gives that

$$B = 2g_x(z_0)\cos(\pi/2 - \alpha\pi) + 2g_y(z_0)\sin(\pi/2 - \alpha\pi).$$

Since (see Figure 6) the unit normal vector $\mathbf{e}$ to the ray $\{re^{i\pi(1-\alpha)} \mid 0 < r < \infty\})$ pointing into the sector S_α defined in (4.18) is

$$\mathbf{e} = \Big(\cos(\pi/2 - \alpha\pi), \sin(\pi/2 - \alpha\pi)\Big),$$

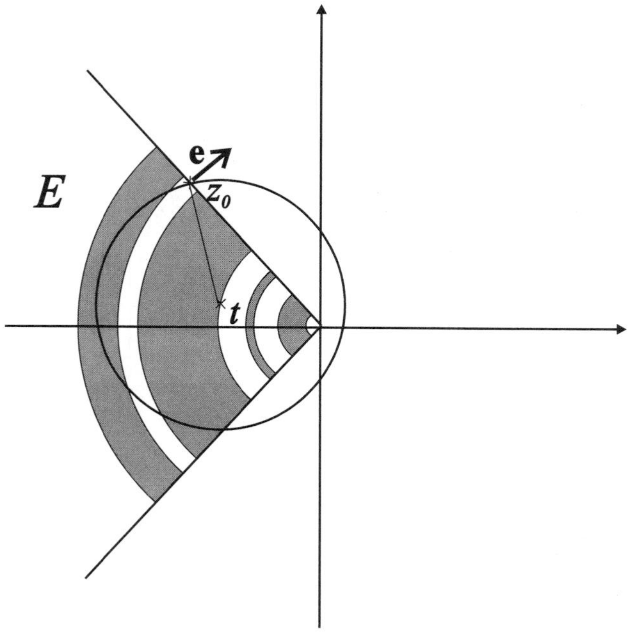

Figure 6:

we can see that $B/2$ is nothing else than the derivative of g at z_0 in the direction of $\mathbf{e}$. Thus, in order to prove (4.19) for sufficiently small ρ it is enough to prove that $B > 0$, i.e. that

$$\frac{\partial g_{\mathbf{C}\setminus E}(z_0)}{\partial \mathbf{e}} = \left.\frac{\partial\left(-U^{\mu_E}(z)\right)}{\partial \mathbf{e}}\right|_{z\,=\,z_0} > 0,$$

where U^{μ_E} denotes the logarithmic potential of the equilibrium measure of E and where we used formula (2.18) for the expression of the Green function in terms of this potential. Thus, we need to show

$$\int_E \left.\frac{\partial \log|z-t|}{\partial \mathbf{e}}\right|_{z\,=\,z_0} d\mu_E(t) > 0, \tag{4.21}$$

which immediately follows from Figure 6 since the function $z \to \log|z-t|$ is increasing in the direction of $\mathbf{e}$. In fact, since z_0 lies on the ray $\{re^{i\pi(1-\alpha)}\,|\,0 < r < \infty\}$ and $\alpha \le 1/2$, the directional derivative of $\log|z-t|$ in the direction of $\mathbf{e}$ at z_0 can be zero only if $\mathbf{e}$ is a tangent vector to the circle $\{z\,|\,|z-t| = |z_0 - t|\}$ (see Figure 6) which means that t is on the same ray $\{re^{i\pi(1-\alpha)}\,|\,0 < r < \infty\}$, or in the case $\alpha = 1/2$ it can also lie on the symmetric ray $\{re^{-i\pi(1-\alpha)}\,|\,0 < r < \infty\}$. Since in all other cases the directional derivative is positive and the set

$$E \setminus \{re^{i\pi(1-\alpha)}\,|\,0 < r < \infty\} \cup \{re^{-i\pi(1-\alpha)}\,|\,0 < r < \infty\}$$

has positive μ_E measure, (4.21) follows.

Note that this very last step heavily used the assumption $0 < \alpha \le 1/2$.

Thus, the theorem has been verified for $\alpha \le 1/2$.

We shall reduce the theorem for $1/2 < \alpha < 1$ to the just proven special case $0 < \alpha \le 1/2$. Recall, that E is of the form depicted in Figure 7. Let k be the

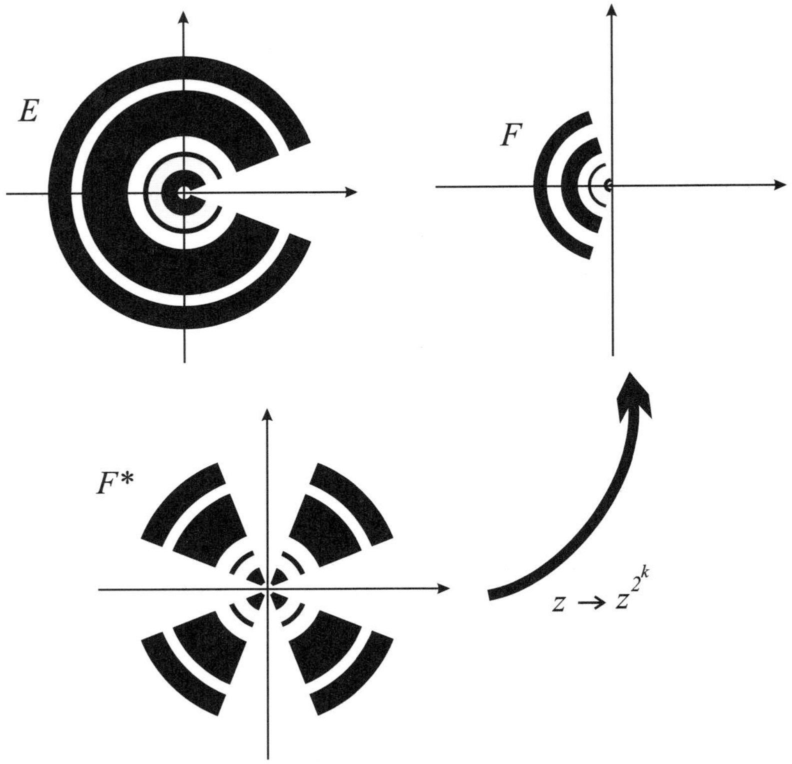

Figure 7:

integer with the property $2^{k-1}(1-\alpha) < 1/2$ but $2^k(1-\alpha) \geq 1/2$, and consider the following set F. Consider those circular arcs on the circles $C_{r^{2^k}}(0)$ that run from the point $r^{2^k}e^{\pi 2^k(1-\alpha)}$ to $r^{2^k}e^{-i\pi 2^k(1-\alpha)}$ and that cross the negative real line. Now let F consist of these arcs for precisely those $0 < r < 1$ for which $C_r(0) \cap E \neq \emptyset$, see Figure 7. The complete inverse image F^* of F under the mapping $z \to z^{2^k}$ is a subset of E: $F^* := \{z \,|\, z^{2^k} \in F\} \subset E$, therefore

$$g_{\mathbf{C}\setminus E}(z) \leq g_{\mathbf{C}\setminus F^*}(z) = \frac{1}{2^k} g_{\mathbf{C}\setminus F}(z^{2^k}),$$

the last equality being immediate from the definition of the Green function. But F is a set for which the theorem has already been proven, since for F the "α value" is $\alpha' = 1 - 2^k(1-\alpha) \leq 1/2$. Thus, on applying the already proven $\alpha' \leq 1/2$ case to $g_{\mathbf{C}\setminus F}$ we get from $1 - \alpha' = 2^k(1-\alpha)$

$$g_{\mathbf{C}\setminus E}(z) \leq C|z^{2^k}|^{1/2 \cdot 2^k(1-\alpha)} \exp\left(D \int_{|z^{2^k}|}^{1} \frac{\Theta_F^2(u)}{u^3} du\right).$$

Furthermore, $F \cap (-\infty, 0]$ is obtained from $E \cap (-\infty, 0]$ by the mapping $t \to -|t|^{2^k}$, therefore Lemma 11.5 of the Appendix shows that

$$\exp\left(D \int_{|z^{2^k}|}^{1} \frac{\Theta_F^2(u)}{u^3} du\right) \leq C_1 \exp\left(D_1 \int_{|z|}^{1} \frac{\Theta_E^2(u)}{u^3} du\right),$$

and the proof is complete.

■

5 Cantor-type sets

Theorem 3.1 shows that in a sense Theorems 2.1–2.3 and their corollaries are sharp. If, however, the set in question has a structure, then this structure may result in much better estimates. As an example consider Cantor type sets. Let $\{\varepsilon_i\}$ be a sequence with $0 \le \varepsilon_i < 1$, and consider the Cantor construction with this sequence. By this we mean that starting from $[0,1]$ first we remove the middle ε_1 part of this interval, then in the second step we remove the middle ε_2 part of both remaining intervals, then in the third step we remove the middle ε_3 part of each remaining 4 intervals, etc. Let us denote the so obtained Cantor type set by $\mathcal{C} = \mathcal{C}(\varepsilon_1, \varepsilon_2, \ldots)$. The classical Cantor set corresponds to the sequence $\varepsilon_j = 1/3$, $j = 1, 2, \ldots$. Since after the n-th step there are 2^n intervals of total length $(1-\varepsilon_1)(1-\varepsilon_2)\cdots(1-\varepsilon_n)$, the set $\mathcal{C}(\varepsilon_1, \varepsilon_2, \ldots)$ is of zero linear measure if and only if $\sum_j \varepsilon_j = \infty$. It is known (see e.g. ([27, V.6.6, Theorem 3])) that $\mathcal{C}(\varepsilon_1, \varepsilon_2, \ldots)$ is of positive capacity if and only if

$$\sum_{j=1}^{\infty} \frac{1}{2^j} \log \frac{1}{1-\varepsilon_j} < \infty.$$

The Green function of $\mathbf{C} \setminus \mathcal{C}(\varepsilon_1, \varepsilon_2, \ldots)$ is in a Lip α class for some positive α if and only if (see ([37]))

$$\sum_{j=1}^{k} \log \frac{1}{1-\varepsilon_j} = O(k). \tag{5.1}$$

We would like to determine when the Green function of $\mathbf{C} \setminus \mathcal{C}(\varepsilon_1, \varepsilon_2, \ldots)$ is of optimal smoothness, i.e. when it belongs to the class Lip 1/2. Corollary 3.3 easily implies that if $\sup_j \varepsilon_j = 1$, then this is not the case, hence in what follows suppose that $0 \le \varepsilon_j \le \varepsilon < 1$ for all j with some fixed $\varepsilon < 1$. Then the sequence $\{t_k\}$ with

$$t_k = \frac{1-\varepsilon_1}{2} \frac{1-\varepsilon_2}{2} \cdots \frac{1-\varepsilon_k}{2}$$

satisfies the condition (3.6), and for the function $\Theta_{\mathcal{C}}(t)$ we have

$$\Theta_{\mathcal{C}}(t_k) = t_k \left(1 - \prod_{j=k+1}^{\infty} (1-\varepsilon_j)\right).$$

Then condition (2.4) for (2.9) takes the form $\sum_k \Theta_{\mathcal{C}}^2(t_k) t_k^{-2} < \infty$, and this is easily seen to be the same as

$$\sum_{k=1}^{\infty} \left(\sum_{j=k+1}^{\infty} \varepsilon_j \right)^2 < \infty.$$

Thus, Theorem 2.2 does not necessarily give the Lip 1/2 property even for Cantor sets of positive measure. But, as the next theorem shows, for all $\mathcal{C}(\varepsilon_1, \varepsilon_2, \ldots)$ of positive measure the Green function $g_{\mathbf{C} \setminus \mathcal{C}}$ is Lip 1/2 smooth. When compared with condition (5.1), this theorem also shows that the Green function of $\mathbf{C} \setminus \mathcal{C}(\varepsilon_1, \varepsilon_2, \ldots)$ to lie in the optimal class Lip 1/2 is a much severe condition than to lie in some Lip α, $\alpha > 0$ class.

Theorem 5.1 *Let $\{\varepsilon_j\}$ be a sequence of numbers from the interval $[0,1)$. The Green function of $\mathbf{C}\setminus\mathcal{C}(\varepsilon_1,\varepsilon_2,\dots)$ is in the class* Lip 1/2 *if and only if $\sum\varepsilon_j^2<\infty$.*

Note that while Theorems 2.1–2.3 give non-trivial estimates only for sets of positive measure (more precisely for sets with the property $\Theta_E(t)=o(t)$), Theorem 5.1 can be applied to sets of zero measure as well. Indeed, we have

Corollary 5.2 *The compact set $\mathcal{C}=\mathcal{C}(1/2,1/3,1/4,\dots)$ is of zero linear measure, but its complement has a* Lip 1/2 *Green function.*

Such a set was constructed by a different method in ([39]).

Theorem 5.1 dramatically illustrates how the structure of a set can influence the behavior of the Green function. Indeed, consider the Cantor construction with the modification that we only omit at the n-th step the middle ε_n part of the leftmost remaining interval (i.e. we omit this time only one interval instead of the 2^{n-1} intervals as in the original construction). Let the so obtained set be $\tilde{\mathcal{C}}(\varepsilon_1,\varepsilon_2,\dots)$. Thus,

$$\tilde{\mathcal{C}}(\varepsilon_1,\varepsilon_2,\dots)=[0,1]\setminus\left(\bigcup_{k=1}^{\infty}(t_k,t_k+\varepsilon_k t_{k-1})\right), \tag{5.2}$$

and clearly $\tilde{\mathcal{C}}$ is a much thicker set than $\mathcal{C}$. Now it easily follows from Theorem 3.1 (c.f. also the proof of Theorem 5.1) that the Green function of $\mathbf{C}\setminus\tilde{\mathcal{C}}(\varepsilon_1,\varepsilon_2,\dots)$ satisfies a Lip 1/2 condition at the origin only if $\sum_j\varepsilon_j^2<\infty$, while Theorem 5.1 shows that this condition is already sufficient for the Lip 1/2 property of the Green function of $\mathbf{C}\setminus\mathcal{C}(\varepsilon_1,\varepsilon_2,\dots)$; and this latter domain has much thinner boundary than $\mathbf{C}\setminus\tilde{\mathcal{C}}(\varepsilon_1,\varepsilon_2,\dots)$.

It is worth recording the following variant of Theorem 5.1.

Theorem 5.3 *Let $\{\varepsilon_j\}$ be a sequence of numbers from the interval $[0,1)$. For the equilibrium measure $\mu_{\mathcal{C}}$ of $\mathcal{C}=\mathcal{C}(\varepsilon_1,\varepsilon_2,\dots)$ we have $\mu_{\mathcal{C}}([0,\delta])=O(\sqrt{\delta})$ if and only if $\sum\varepsilon_j^2<\infty$.*

Although this follows from Theorem 5.1, the sufficiency part in the proof of Theorem 5.1 goes through the proof of Theorem 5.3. Thus, the sufficiency part of Theorem 5.3 will be proven during the verification of the sufficiency of the condition $\sum_i\varepsilon_i^2<\infty$ in Theorem 5.1, Part I. Actually, in the proof of Theorem 5.1 we shall verify the somewhat more general inequality: if $\sum\varepsilon_j^2<\infty$ then $\mu_{\mathcal{C}}([a,a+\delta])=O(\sqrt{\delta})$ uniformly in a.

The necessity of the condition $\sum_i\varepsilon_i^2<\infty$ in Theorem 5.3 is an immediate consequence of the necessity of the same condition in Theorem 5.1, which in turn is a simple consequence of Corollary 3.3 (see the beginning of the proof of Theorem 5.1 below).

In closing this chapter let us remark that the first result on Hölder continuity of Cantor type sets appeared in [14], while the complete characterization (5.1) was given in [37].

5.1 Preliminaries for Cantor sets

Let $\{\varepsilon_i\}$ be a sequence with $0\le\varepsilon_i<1$, and consider the Cantor construction with this sequence. Let us denote by $\mathcal{C}_n(\varepsilon_1,\varepsilon_2,\dots)$ the set that we obtain after n steps during the construction, and let

$$t_n = \frac{1-\varepsilon_1}{2}\frac{1-\varepsilon_2}{2}\cdots\frac{1-\varepsilon_n}{2} \tag{5.3}$$

and

$$t_n^* = \frac{1-\varepsilon_1}{2}\frac{1-\varepsilon_2}{2}\cdots\frac{1-\varepsilon_{n-1}}{2}\frac{1+\varepsilon_n}{2}. \tag{5.4}$$

Then $\mathcal{C}_n$ consists of 2^n intervals of length t_n, and if one of these intervals is $[a, a+t_n]$, then from this interval at the next step we are omitting the interval $(a+t_{n+1}, a+t_{n+1}^*)$ of length $\varepsilon_{n+1}t_n$ keeping the two intervals $[a, a+t_{n+1}]$ and $[a+t_{n+1}^*, t_n]$.

The set $\mathcal{C}_n(\varepsilon_1, \varepsilon_2, \ldots)$ is independent of $\varepsilon_{n+1}, \varepsilon_{n+2}, \ldots$, but for simplicity we shall keep these dummy parameters, as well. Actually, $\mathcal{C}_n$ is a Cantor set on its own right, namely $\mathcal{C}_n = \mathcal{C}(\varepsilon_1, \ldots, \varepsilon_n, 0, 0, \ldots)$. We shall also need the Cantor construction on an interval I, in that case we write $\mathcal{C}_n(I; \varepsilon_1, \varepsilon_2, \ldots)$ for the corresponding sets. Note that if $I = [a, a+t_k]$ is a subinterval of $\mathcal{C}_k$, then $I \cap \mathcal{C}(\varepsilon_1, \varepsilon_2, \ldots)$ is the same as $\mathcal{C}(I; \varepsilon_{k+1}, \varepsilon_{k+2}, \ldots) = a + t_k\mathcal{C}(\varepsilon_{k+1}, \varepsilon_{k+2}, \ldots)$, furthermore, for any $n \geq k$ we have $I \cap \mathcal{C}_n(\varepsilon_1, \varepsilon_2, \ldots) = \mathcal{C}_{n-k}(I; \varepsilon_{k+1}, \varepsilon_{k+2}, \ldots)$.

Next we estimate the capacity of $\mathcal{C}$. We shall work with sequences $\{\varepsilon_i\}$ for which there is an $\varepsilon < 1$ with $\varepsilon_i \leq \varepsilon$. By [27, Sec. V.6.6], if $0 < q \leq \varepsilon_i \leq \varepsilon < 1$, then

$$\operatorname{cap}(\mathcal{C}) \geq \frac{1}{2}q\prod_{i=1}^{\infty}(1-\varepsilon_i)^{1/2^i} \geq \frac{q(1-\varepsilon)}{2}. \tag{5.5}$$

Next we drop the assumption $\varepsilon_i \geq q$. In fact, let $S(\varepsilon_i)$ be the operation applied to a set of intervals Σ that omits the middle $\varepsilon_i|I|$ of every interval I in Σ. Then the Cantor construction is nothing else than repeated application of the operations $S(\varepsilon_1)$, $S(\varepsilon_2)$ (starting from $\Sigma = [0,1]$). Let

$$\varepsilon_i^* = 1 - \frac{(1-\varepsilon_{2i-1})(1-\varepsilon_{2i})}{2}.$$

Note that if we apply $S(\varepsilon_{2i})S(\varepsilon_{2i-1})$ to an interval I, then we obtain four intervals I_1, I_2, I_3, I_4 of length $(1-\varepsilon_{2i-1})(1-\varepsilon_{2i})|I|/4$, and if we apply $S(\varepsilon_i^*)$ to I, then we obtain two intervals of length $(1-\varepsilon_{2i-1})(1-\varepsilon_{2i})|I|/4$, namely the leftmost and rightmost ones of I_1, I_2, I_3, I_4. Thus, $S(\varepsilon_i^*)\Sigma \subset S(\varepsilon_{2i-1})S(\varepsilon_{2i})\Sigma$, and repeated application of this shows that $\mathcal{C}(\varepsilon_1^*, \varepsilon_2^*, \ldots) \subset \mathcal{C}(\varepsilon_1, \varepsilon_2, \ldots)$. But for the numbers ε_i^* we have

$$\frac{1}{2} \leq \varepsilon_i^* \leq 1 - \frac{1}{2}(1-\varepsilon)^2,$$

and so (5.5) can be applied with $q = 1/2$ and $1-(1-\varepsilon)^2/2$ instead of ε. Thus we have proved that if $\varepsilon_i \leq \varepsilon$, then

$$\operatorname{cap}(\mathcal{C}) \geq \frac{(1-\varepsilon)^2}{8}. \tag{5.6}$$

In proving Theorem 5.1 concerning Cantor sets we shall need several preliminary results. In what follows we shall always assume that there is an $\varepsilon < 1$ such that $\varepsilon_i \leq \varepsilon$ is satisfied for all i. Then, in particular, (5.6) holds.

Lemma 5.4 *If $E_1, E_2 \subset [-1,1]$, are compact sets, then*

$$\max_{i=1,2}\operatorname{cap}(E_i) \geq \frac{\operatorname{cap}(E_1 \cup E_2)^2}{2}.$$

Proof. It is known ([29, Theorem 5.1.4]) that

$$\frac{1}{\log(2/\mathrm{cap}(E_1\cup E_2))}\le\frac{1}{\log(2/\mathrm{cap}(E_1))}+\frac{1}{\log(2/\mathrm{cap}(E_2))},$$

and so one of the terms on the right hand side is at least half of the fraction that is on the left hand side. This is exactly the statement in a different form.

■

Lemma 5.5 *If* $-1<A<1$, $E\subset[-1,A]$ *is a compact set and* $\mathrm{cap}(E)\ge\alpha$, *then*

$$\mathrm{cap}(E\setminus(A-\alpha^2,A])\ge\frac{\alpha^2}{2}.$$

Proof. By the previous lemma we have

$$\max\{\mathrm{cap}(E\setminus(A-\alpha^2,A]),\mathrm{cap}([A-\alpha^2,A])\}\ge\frac{\alpha^2}{2}.$$

But $\mathrm{cap}([A-\alpha^2,A])=\alpha^2/4$, hence the statement follows.

■

Lemma 5.6 *Let* E_1 *and* E_2 *be two regular compact subsets of* $[-1,1]$ *in such a way that* E_1 *lies to the left of* E_2, *and let* $E=E_1\cup E_2$. *If* $\mathrm{cap}(E_1)\ge\alpha$, *then*

$$\mu_E(E_1)\ge\frac{\alpha^2}{8\log 2/\alpha^2}.$$

Proof. With $A=\min E_2$ we have $E_1\subset[-1,A]$ and $E_2\subset[A,1]$. Replace E_1 by $E_1^*=E_1\setminus(A-\alpha^2,A]$ and E by $E^*=E_1^*\cup E_2$. Then $\mathrm{dist}(E_1^*,E_2)\ge\alpha^2$, and (see the preceding lemma) $\mathrm{cap}(E_1^*)\ge\alpha^2/2$. Furthermore, μ_{E^*} is the balayage measure of μ_E onto E^*, which implies that $\mu_{E^*}(E_2)\ge\mu_E(E_2)$, and so $\mu_{E^*}(E_1^*)\le\mu_E(E_1)$. Therefore it is enough to estimate $\tau:=\mu_{E^*}(E_1^*)$ from below. For simplicity let us write μ^* instead of μ_{E^*}, and let μ_1^* and μ_2^* be the restriction of μ^* to E_1^* and E_2, respectively.

For $t\in[A,1]$ and $x\in E_1^*$ we have

$$\log\frac{t-A}{t-x}\le\log\frac{1-A}{1-x}=\log\left(1-\frac{A-x}{1-x}\right)\le-\frac{A-x}{1-x}\le-\frac{\alpha^2}{2},$$

therefore

$$\begin{aligned}U^{\mu_2^*}(x)&=U^{\mu_2^*}(A)+\int_{E_2}\log\frac{t-A}{t-x}d\mu_2^*(t)\\&\le U^{\mu_2^*}(A)-\frac{\alpha^2}{2}(1-\tau)\\&=U^{\mu^*}(A)-\frac{\alpha^2}{2}(1-\tau)-U^{\mu_1^*}(A)\\&\le U^{\mu^*}(A)-\frac{\alpha^2}{2}(1-\tau)+\tau\log 2,\end{aligned}$$

where, in the last step we used that $|A-t| \le 2$ for $t \in [-1,1]$.

Now if $\alpha^2/8 \le \tau$, then the statement automatically follows (note that necessarily we have $\alpha \le \mathrm{cap}([-1,1]) = 1/2$). If, however, $\tau \le \alpha^2/8$, then $\tau < 1/2$, and the preceding inequality yields

$$U^{\mu_2^*}(x) \le U^{\mu^*}(A) - \frac{\alpha^2}{4} + \tau \le U^{\mu^*}(A) - \frac{\alpha^2}{8}.$$

The regularity of the sets imply that both $U^{\mu^*}(x)$ and $U^{\mu^*}(A)$ equal to the value $\log 1/\mathrm{cap}(E^*)$, hence together with the preceding inequality we must have

$$U^{\mu_1^*}(x) \ge \frac{\alpha^2}{8},$$

which is the same as

$$U^{\mu_1^*/\tau}(x) \ge \frac{\alpha^2}{8\tau}. \tag{5.7}$$

Here the measure μ_1^*/τ has total mass 1 and is supported on E_1^*, so (see (2.16)) there is an $x \in E_1^*$ for which

$$U^{\mu_1^*/\tau}(x) \le \log \frac{1}{\mathrm{cap}(E_1^*)}.$$

Thus, since (5.7) holds for all $x \in E_1^*$, we must have

$$\mathrm{cap}(E_1^*) \le e^{-\alpha^2/8\tau},$$

and so $\mathrm{cap}(E_1^*) \ge \alpha^2/2$ implies the inequality

$$\frac{\alpha^2}{2} \le e^{-\alpha^2/8\tau},$$

which is the same as

$$\tau \ge \frac{\alpha^2}{8\log(2/\alpha^2)}.$$

■

Lemma 5.7 *Let $L > 2$ and $\beta > 0$ be fixed, and suppose that for $k = 0, 1, \dots, N$ we have for the regular compact sets F_k the relations*

$$\mathrm{cap}(F_k) \ge \beta L^k, \tag{5.8}$$

and $F_k \subset [-L^{k+1}, -2L^k]$ or $F_k \subset [2L^k, L^{k+1}]$. Then for $k = 0, 1, \dots N$

$$\mu_{E_k}(E_0) \le \rho^k, \tag{5.9}$$

where

$$E_k = \bigcup_{j=0}^{k} F_j,$$

and

$$\rho = 1 - \frac{\beta^2}{24L^2\log(2L^2/\beta^2)} \tag{5.10}$$

depends only on β and L.

If (5.8) holds except for K indices between 0 and N, then (5.9) takes the form

$$\mu_{E_k}(E_0) \le \rho^{k-K}. \tag{5.11}$$

Proof. Let μ_k be μ_{E_k}. The claim is clear for $k = 0$. We use induction on k, and suppose that we already know $\mu_k(E_0) \leq \rho^k$. Let, for example, $F_{k+1} \subset [-L^{k+2}, -2L^{k+1}]$. The measure μ_k is the balayage of μ_{k+1} onto E_k, so

$$\mu_k = \mu_{k+1}\Big|_{E_k} + \overline{\mu_{k+1}\Big|_{F_{k+1}}}, \tag{5.12}$$

where $\overline{\nu}$ denotes the balayage of ν onto E_k. We have

$$\overline{\mu_{k+1}\Big|_{F_{k+1}}} = \int_{F_{k+1}} \overline{\delta_a} d\mu_{k+1}(a).$$

Since $\mathrm{dist}(E_k, F_{k+1}) \geq L^{k+1} \geq \mathrm{diam}(E_k)/2$, it follows from $\overline{\delta_\infty} = \mu_k$ and from Harnack's inequality (2.31) that for $a \in F_{k+1}$ we have $\mu_k/3 \leq \overline{\delta_a} \leq 3\mu_k$. On applying Lemma 5.6 to the sets F_{k+1}/L^{k+2} and E_{k+1}/L^{k+2} we can conclude that

$$\overline{\mu_{k+1}\Big|_{F_{k+1}}} \geq \frac{\mu_k}{3}\mu_{k+1}(F_{k+1}) \geq \frac{1}{3}\frac{(\beta/L)^2}{8\log(2L^2/\beta^2)}\mu_k,$$

which, together with (5.12) gives

$$\mu_{k+1}(E_0) + \frac{1}{3}\frac{(\beta/L)^2}{8\log(2L^2/\beta^2)}\mu_k(E_0) \leq \mu_k(E_0),$$

that is

$$\mu_{k+1}(E_0) \leq \left(1 - \frac{1}{3}\frac{(\beta/L)^2}{8\log(2L^2/\beta^2)}\right)\mu_k(E_0),$$

and $\mu_{k+1}(E_0) \leq \rho^{k+1}$ follows from the induction hypothesis.

The proof of the last statement is the same if we remark that we have $\mu_{k+1}(E_0) \leq \mu_k(E_0)$ in all cases.

■

Lemma 5.8 *Suppose that $\varepsilon < 1$ and $\{\varepsilon_i\}$ is a sequence with $\varepsilon_i \leq \varepsilon$. Let $a \in \mathcal{C}_m \setminus \mathcal{C}_{m+1}$ lie in a contiguous interval to $\mathcal{C} = \mathcal{C}(\varepsilon_1, \varepsilon_2, \ldots)$ omitted at the $(m+1)$-st step in the Cantor construction, and let*

$$\frac{(1-\varepsilon)^2}{4} \leq r \leq 1. \tag{5.13}$$

Then with some constants $\rho < 1$ and B that depend only on ε we have

$$\omega\Big(a, C_r(a), (\Delta_r(a) \setminus \mathcal{C}) \cup [a - 2t_m, a + 2t_m]\Big) \leq B\rho^m, \tag{5.14}$$

where

$$t_m = \frac{1-\varepsilon_1}{2}\frac{1-\varepsilon_2}{2}\cdots\frac{1-\varepsilon_m}{2},$$

is the number from (5.3).

Furthermore, for any $n \geq m$

$$\begin{aligned} &\omega(a, C_r(a), \Delta_r(a) \setminus \mathcal{C}_n) \\ &\quad\leq\; 3B\rho^m\omega(1/2, C_{1/2}(1/2), \Delta_{1/2}(1/2) \setminus \mathcal{C}_{n-m}(\varepsilon_{m+1}, \varepsilon_{m+2}, \ldots)). \end{aligned} \tag{5.15}$$

Of course, the lower bound in (5.13) is only for definiteness, any positive fixed constant could replace the left hand side.

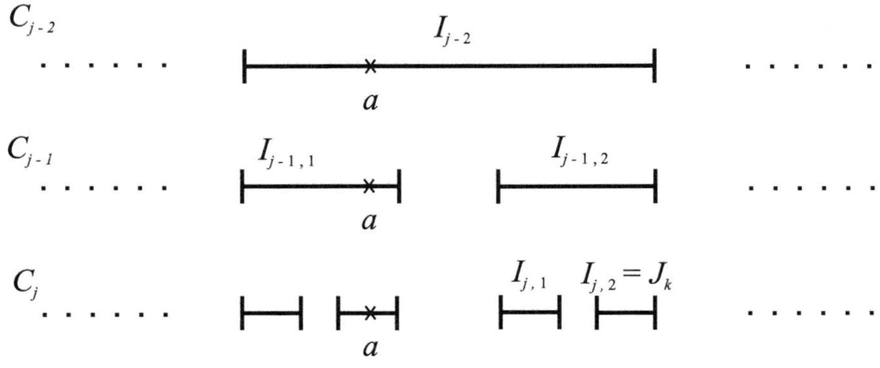

Figure 8:

Proof. Let $L = 256/(1-\varepsilon)^5$. We show that we can transform the problem to Lemma 5.7 with this L. Let N be the largest number for which

$$4t_m \le L^{-(N+1)}$$

is satisfied. If there is no such N or $N = 0$, then

$$t_m \ge \frac{1}{4} L^{-2} > \frac{(1-\varepsilon)^{10}}{2^{18}},$$

so $m < 18$, and (5.14) is true with $B = 2^{17}$, with any $\rho \ge 1/2$. Thus, we may suppose $N \ge 1$.

Consider a $1 \le k \le N$, and let $j = j_k$ be the largest index with $2L^{-k-1} \le t_j$. Then

$$\frac{2}{L^{k+1}} \le t_j \le \frac{4}{1-\varepsilon} \frac{1}{L^{k+1}}. \tag{5.16}$$

The point a belongs to some subinterval I_{j-2} of $\mathcal{C}_{j-2}$ (note that

$$t_j > 8t_m, \tag{5.17}$$

so $j < m$ and a does belong to the set $\mathcal{C}_j$). The interval I_{j-2} is the union of two subintervals of $\mathcal{C}_{j-1}$, say $I_{j-2} = I_{j-1,1} \cup I_{j-1,2}$, where $I_{j-1,1}$ lies to the left of $I_{j-1,2}$ (see Figure 8). The point a belongs to one of these subintervals, say to $I_{j-1,1}$. Now the other one is the union of two subintervals of $\mathcal{C}_j$, say $I_{j-1,2} = I_{j,1} \cup I_{j,2}$, where $I_{j,1}$ lies to the left of $I_{j,2}$. We shall work in what follows with $I_{j,2}$, and let us call it J_k (eventually it has been chosen for the number k).

It is clear from the construction that

$$\operatorname{dist}(a, J_k) \ge |I_{j,1}| = t_j, \qquad \operatorname{diam}(\{\{a\} \cup J_k\}) \le |I_{j-2}| = t_{j-2} \le \frac{4}{(1-\varepsilon)^2} t_j. \tag{5.18}$$

The set $J_k \cap \mathcal{C}$ is a scaled version of $\mathcal{C}(\varepsilon_{j+1}, \varepsilon_{j+2}, \ldots)$ with scaling factor t_j, hence it follows from (5.6) that

$$\operatorname{cap}(J_k \cap \mathcal{C}) \ge t_j (1-\varepsilon)^2/8. \tag{5.19}$$

The interval J_k lies in

$$\left[a + t_j, a + \frac{4}{(1-\varepsilon)^2} t_j\right] \subset \left[a, a + \frac{r}{2}\right] \tag{5.20}$$

because

$$\frac{4}{(1-\varepsilon)^2} t_j \le \frac{16}{(1-\varepsilon)^3} \frac{1}{L^{k+1}} \le \frac{r}{2}.$$

Consider now the mapping

$$T(z) = \frac{1}{2}\left(\frac{z-a}{r} + \frac{r}{z-a}\right),$$

that maps $C_r(a)$ into $[-1, 1]$, and let F_k be the image of $J_k \cap \mathcal{C}$ under this mapping. For $x \in J_k$ we have (see (5.16) – (5.18) and (5.13))

$$\frac{(1-\varepsilon)^5}{64} L^{k+1} \le \frac{(1-\varepsilon)^2}{4t_{j-2}} \le \frac{r}{x-a} \le \frac{1}{t_j} \le \frac{1}{2} L^{k+1}$$

and hence

$$0 \le \frac{x-a}{r} \le \frac{1}{2}. \tag{5.21}$$

These give

$$2L^k \le \frac{(1-\varepsilon)^5}{128} L^{k+1} \le T(x) \le \frac{1}{2}\left(\frac{L^{k+1}}{2} + 1\right) \le L^{k+1}.$$

Thus, $F_k \subset [2L^k, L^{k+1}]$. For $x \in J_k$ we can write

$$\begin{aligned} |T'(x)| &= \left|\frac{1}{2}\left(\frac{1}{r} - \frac{r}{(x-a)^2}\right)\right| \ge \frac{r}{4(x-a)^2} \ge \frac{r}{4}\frac{1}{t_{j-2}^2} \\ &\ge \frac{r}{4}\frac{(1-\varepsilon)^4}{16}\frac{1}{t_j^2} \ge \frac{(1-\varepsilon)^6}{256}\frac{1}{t_j^2} \ge \frac{(1-\varepsilon)^7 L^{k+1}}{1024 t_j} = \frac{(1-\varepsilon)^2}{4t_j} L^k, \end{aligned}$$

where at the first inequality we used that

$$\frac{r}{(x-a)^2} \ge \frac{2}{r}$$

(see (5.21)), and where at the other inequalities we used (5.16) and (5.18). Thus, on J_k the mapping T is expansive with expansion factor $\ge (1-\varepsilon)^2 L^k / 4t_j$, which implies (see (5.19))

$$\operatorname{cap}(F_k) \ge \frac{(1-\varepsilon)^2}{4t_j} L^k \operatorname{cap}(J_k \cap \mathcal{C}) \ge \frac{(1-\varepsilon)^4}{32} L^k.$$

This is true for $k = 1, \ldots, N$.

Now as we have already seen in (5.18) and (5.17)

$$\operatorname{dist}(a, J_k) \ge t_j \ge 8t_m,$$

which implies

$$\bigcup_{k=1}^{N} (J_k \cap \mathcal{C}) \subseteq \mathcal{C} \setminus [a - 2t_m, a + 2t_m],$$

and hence

$$\begin{aligned}&\omega\Big(a, C_r(a), (\Delta_r(a) \setminus \mathcal{C}) \cup [a - 2t_m, a + 2t_m]\Big)\\ \le\ &\omega\Big(a, C_r(a), \Delta_r(a) \setminus (\cup_{k=1}^N (J_k \cap \mathcal{C}))\Big) = \omega\Big(\infty, [-1,1], \mathbf{C} \setminus \cup_{k=0}^N F_k\Big),\end{aligned}$$

where we set $F_0 = [-1, 1]$. But the last harmonic measure is the same as the mass of the equilibrium measure of $\cup_{k=0}^N F_k$ on $[-1, 1]$, hence we obtain from the preceding inequality and from Lemma 5.7 that

$$\omega\Big(a, C_r(a), (\Delta_r(a) \setminus \mathcal{C}) \cup [a - 2t_m, a + 2t_m]\Big) \le \rho_1^N,$$

where ρ_1 is the number (5.10) from that lemma for the parameters L and $\beta = (1 - \varepsilon)^4/32$. Here by the choice of the number N we have

$$\frac{1}{L^N} < 4L^2 t_m \le \frac{2^{18}}{(1-\varepsilon)^{10}} \left(\frac{1}{2}\right)^m,$$

and so $\rho_1^N \le B\rho^m$ with

$$\rho = \left(\frac{1}{2}\right)^{\log(1/\rho_1)/\log L}$$

and

$$B = \frac{2^{18}}{(1-\varepsilon)^{10}}.$$

All these imply (5.14), and the proof of the first part of the lemma is complete.

The inequality (5.15) will be deduced from (5.14). Let I be the subinterval of $\mathcal{C}_m$ containing a and let Δ^* be the disk with diameter I. Clearly

$$\begin{aligned}\omega\Big(a, C_r(a), \Delta_r(a) \setminus \mathcal{C}_n\Big) \ &\le\ \omega\Big(a, \partial\Delta^*, \Delta^* \setminus \mathcal{C}_n\Big) \times \\ &\times\ \max_{b \in \partial\Delta^*} \omega\Big(b, C_r(a), \Delta_r(a) \setminus \mathcal{C}_n\Big).\end{aligned} \tag{5.22}$$

Here the first factor on the right is the same as $\omega(a', C_{1/2}(1/2), \Delta_{1/2}(1/2) \setminus \mathcal{C}')$, where a' and $\mathcal{C}'$ are the images of a and $\mathcal{C}_n \cap I$ under the linear mapping that maps I into $[0, 1]$. Since $\mathcal{C}_n \cap I$ is $\mathcal{C}_{n-m}(I; \varepsilon_{m+1}, \varepsilon_{m+2}, \ldots)$, we have

$$\mathcal{C}' = \mathcal{C}_{n-m}(\varepsilon_{m+1}, \varepsilon_{m+2}, \ldots)$$

and $a' \in ((1 - \varepsilon_{m+1})/2, (1 + \varepsilon_{m+1})/2)$, hence it follows from Lemma 2.4 that

$$\omega\Big(a, \partial\Delta^*, \Delta^* \setminus \mathcal{C}_n\Big) \le \omega\Big(1/2, C_{1/2}(1/2), \Delta_{1/2}(1/2) \setminus \mathcal{C}_{n-m}(\varepsilon_{m+1}, \varepsilon_{m+2}, \ldots)\Big).$$

In estimating the supremum on the right of (5.22) we use (5.14). First of all

$$\omega\Big(b, C_r(a), \Delta_r(a) \setminus \mathcal{C}_n\Big) \le \omega\Big(b, C_r(a), (\Delta_r(a) \setminus \mathcal{C}) \cup [a - 2t_m, a + 2t_m]\Big).$$

Here the function

$$\omega\Big(z, C_r(a), (\Delta_r(a) \setminus \mathcal{C}) \cup [a - 2t_m, a + 2t_m]\Big)$$

is harmonic in $(\Delta_r(a) \setminus \mathcal{C}) \cup [a - 2t_m, a + 2t_m]$ which contains the disk $\Delta_{2t_m}(a)$, hence Harnack's inequality (2.30) gives for $b \in \partial\Delta^* \subset \Delta_{t_m}(a)$

$$\omega\Big(b, C_r(a), (\Delta_r(a) \setminus \mathcal{C}) \cup [a - 2t_m, a + 2t_m]\Big) \\ \leq 3\omega\Big(a, C_r(a), (\Delta_r(a) \setminus \mathcal{C}) \cup [a - 2t_m, a + 2t_m]\Big).$$

On applying (5.14) we can see that this is at most $3B\rho^m$, and the proof of (5.15) is complete.

■

5.2 Proof of Theorems 5.1 and 5.3

For simpler notation in this section we use the abbreviations $C_0 = C_{1/2}(1/2)$ and $\Delta_0 = \Delta_{1/2}(1/2)$ for the circle an disk with diameter $[0, 1]$. In the proof of Theorem 5.1 we shall need a good bound for the harmonic measure $\omega(x, C_0, \Delta_0 \setminus \mathcal{C})$, at least in the middle interval $((1 - \varepsilon)/2, (1 + \varepsilon_1)/2)$ that is omitted at the very first step during the Cantor construction.

Theorem 5.9 *Let $\varepsilon < 1$. Then there are constants $\sigma < 1$ and D, both depending only on ε, such that if $\{\varepsilon_i\}$ is any sequence with $\varepsilon_i \leq \varepsilon$ and $x \in [(1 - \varepsilon_1)/2, (1 + \varepsilon_1)/2)]$, then*

$$\omega(x, C_0, \Delta_0 \setminus \mathcal{C}(\varepsilon_1, \varepsilon_2, \ldots)) \leq D(\varepsilon_1 + \sum_{j>1} \varepsilon_j^2 \sigma^{j-1}). \tag{5.23}$$

Although we shall need only this lemma, the following consequence will be verified during the following proofs. We state this consequence here for completeness regarding the estimate of $\omega(x, C_0, \Delta_0 \setminus \mathcal{C})$ for any $x \in [0,1] \setminus \mathcal{C}$. In fact, let $x \in [0,1] \setminus \mathcal{C}$, say x lies in $[1, 1/2]$. Then, using the numbers t_n from (5.3), we can assume that x lies in $[0, t_1]$ for otherwise we can apply the theorem. Let k be the index with $x \in [t_{k+1}, t_k]$, and let $m \geq 1$ be the number for which $x \in \mathcal{C}_{k+m} \setminus \mathcal{C}_{k+m+1}$, i.e. the contiguous interval to $\mathcal{C}$ containing x is omitted at the $(k + m + 1)$-st step. Note that this interval is of length $\varepsilon_{k+m+1} t_{k+m}$. Now Lemma 5.8, (5.15) and Theorem 5.9 imply that with some constants $\sigma < 1$ and D that depend only on the bound ε for the numbers ε_i we have

$$\omega(x, C_0, \Delta_0 \setminus \mathcal{C}_n(\varepsilon_1, \varepsilon_2, \ldots)) \leq D\sigma^m(\varepsilon_{k+m+1} + \sum_{j>k+m+1} \varepsilon_j^2 \sigma^{j-(k+m+1)}). \tag{5.24}$$

Proof of Theorem 5.9

In this proof we shall frequently use the numbers

$$t_n = \frac{1 - \varepsilon_1}{2} \frac{1 - \varepsilon_2}{2} \cdots \frac{1 - \varepsilon_n}{2}$$

and

$$t_n^* = \frac{1 - \varepsilon_1}{2} \frac{1 - \varepsilon_2}{2} \cdots \frac{1 - \varepsilon_{n-1}}{2} \frac{1 + \varepsilon_n}{2}$$

that were defined in (5.3) and (5.4).

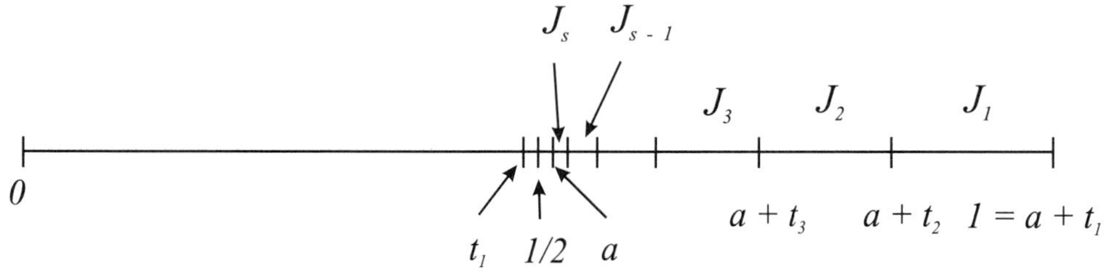

Figure 9:

Let $\varepsilon < 1$ and consider all sequences $\{\varepsilon_i\}$ with $\varepsilon_i \le \varepsilon$. Let $\rho = \rho(\varepsilon) < 1$ be the constant in Lemma 5.8 for this ε, and set $\sigma = \rho^{1/4}$. We claim that with this σ the assertion of the theorem is true for some constant D. To this end let us denote by $D_n = D_n(\varepsilon)$ the smallest constant for which

$$\omega(x, C_0, \Delta_0 \setminus \mathcal{C}_n(\varepsilon_1, \varepsilon_2, \ldots)) \le D_n(\varepsilon_1 + \sum_{j>1} \varepsilon_j^2 \sigma^{j-1}) \tag{5.25}$$

is true for all $x \in [(1-\varepsilon_1)/2, (1+\varepsilon_1)/2]$ and for all sequences $\{\varepsilon_i\}$ with $\varepsilon_i \le \varepsilon$, where $\mathcal{C}_n$ denotes the set at the n-th level during the Cantor construction. We claim that $\{D_n\}$ is a bounded sequence, and in the limit we obtain the assertion of the theorem.

As we have seen in Lemma 2.4, because of the symmetry of $\mathcal{C}$ onto 1/2, the harmonic measure $\omega(x, C_0, \Delta_0 \setminus \mathcal{C})$ attains its maximum on the interval $[(1-\varepsilon_1)/2, (1+\varepsilon_1)/2]$ at $x = 1/2$, hence it is enough to estimate this value. We obtain from Lemma 2.5

$$\omega(1/2, C_0, \Delta_0 \setminus [(1-\varepsilon_1)/2, (1+\varepsilon_1)/2]) \le \frac{4}{\pi} 2\varepsilon_1,$$

which gives

$$D_1 \le \frac{8}{\pi}.$$

Next we estimate D_n in terms of D_{n-1}.

We have the representation

$$\mathcal{C}_n(\varepsilon_1, \varepsilon_2, \ldots) = \mathcal{C}_{n-1}([0, (1-\varepsilon_1)/2]; \varepsilon_2, \varepsilon_3, \ldots) \cup \mathcal{C}_{n-1}([(1+\varepsilon_1)/2, 1]; \varepsilon_2, \varepsilon_3, \ldots),$$

and on the right hand side the two sets are symmetric with respect to the point 1/2, therefore we shall consider only the second one. Let s be the index with

$$t_s < \varepsilon_1 \le t_{s-1}, \tag{5.26}$$

and for simplicity let us denote $(1+\varepsilon_1)/2$ by a. The interval $[a, 1]$ is the union of the intervals $J_k = [a + t_{k+1}, a + t_k]$ with $k = 1, \ldots, s-1$ and of the interval $J_s = [a, a + t_s]$ (see Figure 9).

Note that $J_k \cap \mathcal{C}_n$ is nothing else than $\mathcal{C}_{n-k-1}([a + t_{k+1}^*, a + t_k]; \varepsilon_{k+2}, \varepsilon_{k+3}, \ldots)$ (when $k = s$ it is $\mathcal{C}_{n-k}([a, a + t_s]; \varepsilon_{k+1}, \varepsilon_{k+2}, \ldots))$. Its complement relative to J_k

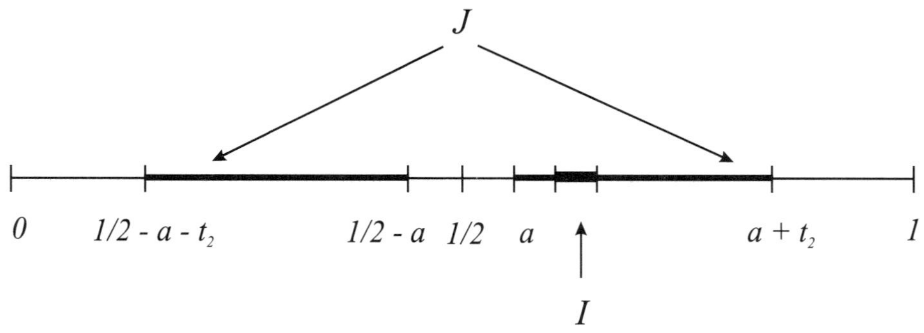

Figure 10:

can be written as

$$J_k \setminus \mathcal{C}_n = \bigcup_{m=0}^{n-k-1} J_k \cap (\mathcal{C}_{k+m} \setminus \mathcal{C}_{k+m+1}), \tag{5.27}$$

and here the set $J_k \cap (\mathcal{C}_{k+m} \setminus \mathcal{C}_{k+m+1})$ consists of intervals of identical length $\varepsilon_{k+m+1}t_{k+m}$, the number of which is 1 if $m = 0$ and 2^{m-1} if $m \geq 1$. Let I be one of these intervals and $x \in I$. If $k \geq 2$ (we shall need the estimate below only for such k's) then $I \subset [1/2, 1-(1-\varepsilon)^2/4]$, so we obtain from Lemma 5.8, (5.15) with $r = (1-\varepsilon)^2/4$ that for $x \in I$

$$\begin{aligned}\omega(x, C_0, \Delta_0 \setminus \mathcal{C}_n) &\leq \omega(x, C_r(x), \Delta_r(x) \setminus \mathcal{C}_n) \\ &\leq 3B\rho^{k+m}\omega(1/2, C_0, \Delta_0 \setminus \mathcal{C}_{n-k-m}(\varepsilon_{k+m+1}, \varepsilon_{k+m+2}, \ldots)).\end{aligned}$$

In view of the definition of the numbers D_i we have

$$\begin{aligned}&\omega(1/2, C_0, \Delta_0 \setminus C_{n-m-k}(\varepsilon_{k+m+1}, \varepsilon_{k+m+2}, \ldots)) \\ &\qquad \leq D_{n-k-m}(\varepsilon_{k+m+1} + \sum_{j>k+m+1} \varepsilon_j^2 \sigma^{j-(k+m+1)}),\end{aligned}$$

and so

$$\omega(x, C_0, \Delta_0 \setminus \mathcal{C}_n) \leq 3B\rho^{m+k} D_{n-1}(\varepsilon_{k+m+1} + \sum_{j>k+m+1} \varepsilon_j^2 \sigma^{j-(k+m+1)}). \tag{5.28}$$

Let J be the set $[a, a+t_2] \cup [1/2-a-t_2, 1/2-a]$ (see Figure 10). Next we estimate for an interval $I \subset J_k$, $k \geq 2$ the value $\omega(1/2, I, \Delta_0 \setminus J)$. This is at most as large as

$$\omega(1/2, I, \mathbf{C} \setminus J) \leq \omega(1/2, I \cup (1/2 - I), \mathbf{C} \setminus J) = \omega(0, [\alpha, \beta], \mathbf{C} \setminus [p, q]),$$

where in the last step we applied the transformation $T(z) = (z-1/2)^2$, and $[\alpha, \beta]$ resp. $[p, q]$, are the images of I resp. J under this transformation (c.f. Figure 11). Thus,

$$p = \left(\frac{\varepsilon_1}{2}\right)^2, \quad \frac{(1-\varepsilon)^4}{16} \leq \left(\frac{\varepsilon_1}{2} + t_2\right)^2 = q \leq \frac{1}{4}$$

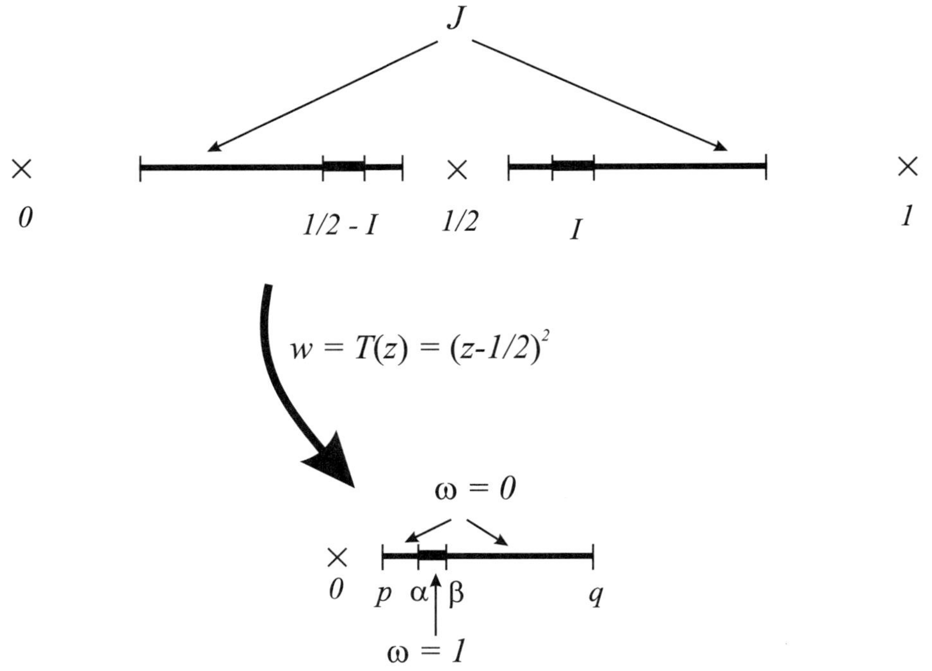

Figure 11:

and

$$[\alpha,\beta]\subseteq T(J_k)=\left[\left(\frac{\varepsilon_1}{2}+t_{k+1}\right)^2,\left(\frac{\varepsilon_1}{2}+t_k\right)^2\right]$$

except when $k=s$, in which case

$$[\alpha,\beta]\subseteq T(J_s)=\left[\left(\frac{\varepsilon_1}{2}\right)^2,\left(\frac{\varepsilon_1}{2}+t_s\right)^2\right].$$

Now $\omega(0,[\alpha,\beta],\mathbf{C}\setminus[p,q])$ equals $\widehat{\delta_0}([\alpha,\beta])$, where $\widehat{\delta_0}$ denotes the balayage of the Dirac delta δ_0 at 0 onto the interval $[p,q]$. Thus, we can apply (2.27) to calculate for $2<k<s$

$$\begin{aligned}\omega(0,[\alpha,\beta],\mathbf{C}\setminus[p,q]) &= \frac{1}{\pi}\int_\alpha^\beta \frac{\sqrt{p}\sqrt{q}}{t\sqrt{t-p}\sqrt{q-t}}dt \\ &\leq B_1\int_\alpha^\beta \frac{\varepsilon_1}{t_{k+1}^2\sqrt{t_{k+1}^2}}dt = B_1\frac{\varepsilon_1(\beta-\alpha)}{t_{k+1}^3}\leq B_2\frac{\varepsilon_1|I|}{t_k^2},\end{aligned} \tag{5.29}$$

where the numbers B_1, B_2 depend only on ε and where we have used that because of $\varepsilon_1\leq t_{s-1}\leq t_k$ we have

$$\beta-\alpha\leq |I|2\left(\frac{\varepsilon_1}{2}+t_k\right)\leq 3|I|t_k.$$

This estimate has to be changed if $k = s$ or $k = 2$ because, e.g. for $k = s$, the interval I can lie arbitrarily close to a so α can be arbitrarily small, while for $k < s$ we have $\alpha \geq t_{k+1}^2$. When $k = s$ we set $I = [a + \gamma\varepsilon_1, a + \gamma\varepsilon_1 + \delta\varepsilon_1]$, where, clearly, $0 \leq \gamma \leq \gamma + \delta \leq 1$, since I lies in the interval J_s which is contained in $[a, a + t_s] \subset [a, a + \varepsilon_1]$ (c.f. (5.26)). Then

$$\alpha = \left[\left(\frac{1}{2} + \gamma\right)\varepsilon_1\right]^2, \qquad \beta = \left[\left(\frac{1}{2} + \gamma + \delta\right)\varepsilon_1\right]^2.$$

We get for the balayage integral in (5.29) the estimate

$$\begin{aligned} &\omega(1/2, I, \Delta_0 \setminus J) &(5.30)\\ &\leq B_3 \frac{\varepsilon_1}{\varepsilon_1^2} \int_\alpha^\beta \frac{dt}{\sqrt{t-p}} = \frac{2B_3}{\varepsilon_1}\left(\sqrt{\beta - p} - \sqrt{\alpha - p}\right) \\ &= \frac{2B_3}{\varepsilon_1} \frac{\beta - \alpha}{\sqrt{\beta - p} + \sqrt{\alpha - p}} = \frac{2B_3}{\varepsilon_1} \frac{\delta(2\gamma + \delta + 1)\varepsilon_1^2}{(\sqrt{(\gamma+\delta)(\gamma+\delta+1)} + \sqrt{\gamma(\gamma+1)})\varepsilon_1} \\ &\leq 2B_3 \frac{\delta \cdot 4}{\sqrt{\gamma + \delta}} = 2B_3 \frac{4|I|}{\varepsilon_1\sqrt{\gamma+\delta}}, \end{aligned}$$

and note that here $\gamma + \delta$ is $1/\varepsilon_1$ times the distance from the right endpoint of I to a.

In a similar manner, if $k = 2$ and τ is the distance from the left endpoint of I to $a + t_2$, which is the right endpoint of J, then

$$\omega(1/2, I, \Delta_0 \setminus J) \leq B_4\varepsilon_1 \int_\alpha^\beta \frac{dt}{\sqrt{q-t}} = 2B_4\varepsilon_1 \frac{\beta - \alpha}{\sqrt{q-\alpha} + \sqrt{q - \beta}} \leq B_5\varepsilon_1 \frac{|I|}{\sqrt{\tau}} \tag{5.31}$$

where again, the constant B_5 depends only on ε.

This completes the estimate of $\omega(1/2, I, \Delta_0 \setminus J)$.

After all these consider the function

$$h(z) = \omega(z, C_0, \Delta_0 \setminus J) + \sum_I \omega(z, I, \Delta_0 \setminus J) \max_{x \in I} \omega(x, C_0, \Delta_0 \setminus \mathcal{C}_n), \tag{5.32}$$

where the summation is extended over all subintervals of $J \setminus \mathcal{C}_n$. This is harmonic in $\Delta_0 \setminus J$, and its boundary values are at least as large as the corresponding values of $\omega(z, C_0, \Delta_0 \setminus \mathcal{C}_n)$ on $\partial(\Delta_0 \setminus J)$. Thus, we get from the maximum principle that

$$\omega(1/2, C_0, \Delta_0 \setminus \mathcal{C}_n) \leq h(1/2). \tag{5.33}$$

For $z = 1/2$ the first term in the definition of $h(z)$ can be bounded as (see Lemma 2.5)

$$\begin{aligned} \omega(1/2, C_0, \Delta_0 \setminus J) &\leq \omega(1/2, C_{\varepsilon_1/2+t_2}(1/2), \Delta_{\varepsilon_1/2+t_2}(1/2) \setminus J) \\ &\leq \frac{4}{\pi t_2}\varepsilon_1 \leq \frac{16}{\pi(1-\varepsilon)^2}\varepsilon_1. \end{aligned} \tag{5.34}$$

The subintervals I in the sum defining $h(z)$ lie symmetrically onto $1/2$, and everything is symmetric onto $1/2$, therefore we shall concentrate only on the subintervals

I lying in $[1/2, 1]$. Let us group the sum $\sum_{I\subset([1/2,1]\cap J)}$ in the definition of $h(1/2)$ as

$$\sum_I = \sum_{k=2}^{s} \sum_{I\subset J_k} =: \sum_{k=2}^{s} \Sigma_k. \tag{5.35}$$

If $2 < k < s$ then it follows from (5.27), the discussion after it, from (5.28) and from (5.29) that Σ_k is bounded by

$$3BD_{n-1} \sum_{m=0}^{n-k-1} 2^m \rho^{k+m} (\varepsilon_{k+m+1} + \sum_{j>k+m+1} \varepsilon_j^2 \sigma^{j-(k+m+1)}) \frac{B_2 \varepsilon_1 (\varepsilon_{k+m+1} t_{k+m})}{t_k^2},$$

and so

$$\Sigma_k \leq 3BB_2 D_{n-1} \frac{\varepsilon_1}{t_k} \sum_{m=0}^{n-k-1} \rho^{k+m} \varepsilon_{k+m+1} (\varepsilon_{k+m+1} + \sum_{j>k+m+1} \varepsilon_j^2 \sigma^{j-(k+m+1)}) \tag{5.36}$$

because $t_{k+m}/t_k \leq 2^{-m}$.

When $k = s$, the situation is somewhat more involved. In this case note that $J_s \cap \mathcal{C}_{s+m}$ consists of 2^m equal subintervals of length t_{s+m}, and so $J_s \cap (\mathcal{C}_{s+m} \setminus \mathcal{C}_{s+m+1})$ consists of 2^m equal subintervals of length $\varepsilon_{s+m+1} t_{s+m}$, where the right endpoint of the l-th subinterval (counted from the left) is at distance $\geq l t_{s+m+1} = (l t_{s+m+1}/\varepsilon_1)\varepsilon_1$ from the point $a = (1+\varepsilon_1)/2$. Therefore, we obtain from (5.27), (5.28) and (5.30) that Σ_s is bounded by

$$\begin{aligned}
&3BD_{n-1} \sum_{m=0}^{n-k-1} \rho^{s+m} (\varepsilon_{s+m+1} + \sum_{j>s+m+1} \varepsilon_j^2 \sigma^{j-(s+m+1)}) \times \\
&\qquad \times 8B_3 \sum_{l=1}^{2^m} \frac{\varepsilon_{s+m+1} t_{s+m}}{\varepsilon_1 \sqrt{l t_{s+m+1}/\varepsilon_1}} \\
&\leq B_6 D_{n-1} \sum_{m=0}^{n-k-1} \rho^{s+m} \varepsilon_{s+m+1} (\varepsilon_{s+m+1} + \sum_{j>s+m+1} \varepsilon_j^2 \sigma^{j-(s+m+1)}) \times \\
&\qquad \times \sqrt{\frac{t_{s+m}}{\varepsilon_1}} \sum_{l=1}^{2^m} \frac{1}{\sqrt{l}} \\
&\leq 2B_6 D_{n-1} \sum_{m=0}^{n-k-1} \rho^{s+m} \varepsilon_{s+m+1} (\varepsilon_{s+m+1} + \sum_{j>s+m+1} \varepsilon_j^2 \sigma^{j-(s+m+1)})
\end{aligned}$$

because

$$\frac{t_{s+m}}{t_{s+m+1}} \leq \frac{2}{1-\varepsilon}$$

and

$$\sqrt{\frac{t_{s+m}}{\varepsilon_1}} \sum_{l=1}^{2^m} \frac{1}{\sqrt{l}} \leq \sqrt{\frac{t_{s+m}}{\varepsilon_1}} 2\sqrt{2^m} \leq 2\sqrt{\frac{t_s}{\varepsilon_1}} \leq 2$$

by the choice of the number s. Thus, the estimate (5.36) is correct for $k = s$, if we replace the constant $3BB_2$ by $2B_6$ (recall that $1 \leq \varepsilon_1/t_s$).

In a similar manner we can use (5.31) to see that Σ_2 is bounded by

$$3BD_{n-1}\sum_{m=0}^{n-3}\rho^{2+m}(\varepsilon_{2+m+1}+\sum_{j>2+m+1}\varepsilon_j^2\sigma^{j-(2+m+1)})\sum_{l=1}^{2^m}\frac{B_5\varepsilon_1\varepsilon_{2+m+1}t_{2+m}}{\sqrt{lt_{2+m+1}}}$$
$$\leq\ 3BB_5D_{n-1}\sqrt{\frac{2}{1-\varepsilon}}\varepsilon_1\sum_{m=0}^{n-3}\rho^{2+m}\varepsilon_{2+m+1}(\varepsilon_{2+m+1}+\sum_{j>2+m+1}\varepsilon_j^2\sigma^{j-(2+m+1)})$$

where we used that

$$\sum_{l=1}^{2^m}\frac{t_{2+m}}{\sqrt{lt_{2+m+1}}}\ \leq\ \sqrt{\frac{2}{1-\varepsilon}}\sqrt{t_{2+m}}\sum_{l=1}^{2^m}\frac{1}{\sqrt{l}}$$
$$\leq\ \sqrt{\frac{2}{1-\varepsilon}}\sqrt{t_{2+m}}2\sqrt{2^m}\leq\sqrt{\frac{2}{1-\varepsilon}}.$$

This shows that (5.36) is correct for $k=2$, as well, if we replace the constant $3BB_2$ by $3BB_5$.

Summing up all for $k=2,\dots,s$ the above inequalities we finally obtain with a constant B_7 that depends only on ε that (5.35) in (5.32) can be bounded as

$$\sum_{k=2}^{s}\Sigma_k\leq$$
$$\leq\ B_7D_{n-1}\sum_{k=2}^{s}\frac{\varepsilon_1}{t_k}\sum_{m=0}^{n-k-1}\rho^{k+m}\varepsilon_{k+m+1}(\varepsilon_{k+m+1}+\sum_{j>k+m+1}\varepsilon_j^2\sigma^{j-(k+m+1)})$$
$$\leq\ B_7D_{n-1}\Bigg[\sum_{m=2}^{s}\varepsilon_{m+1}(\varepsilon_{m+1}+\sum_{j>m+1}\varepsilon_j^2\sigma^{j-(m+1)})\varepsilon_1\rho^m\sum_{k=2}^{m}\frac{1}{t_k}+$$
$$+\sum_{m=s+1}^{\infty}\varepsilon_{m+1}(\varepsilon_{m+1}+\sum_{j>m+1}\varepsilon_j^2\sigma^{j-(m+1)})\varepsilon_1\rho^m\sum_{k=2}^{s}\frac{1}{t_k}\Bigg].$$

Here for $m<s$

$$\varepsilon_1\sum_{k=2}^{m}\frac{1}{t_k}\leq\varepsilon_1\frac{1}{t_m}\sum_{l}\frac{1}{2^l}\leq\varepsilon_1\frac{2}{t_m}\leq2\frac{t_{s-1}}{t_m}\leq2\cdot2^{-(s-1-m)},$$

while for $m=s$

$$\varepsilon_1\sum_{k=2}^{s}\frac{1}{t_k}\leq\varepsilon_1\frac{2}{t_s}\leq2\frac{t_{s-1}}{t_s}\leq\frac{4}{1-\varepsilon}.$$

Hence, since $1/2<\rho$, we can infer

$$\sum_{k=2}^{s}\Sigma_k\leq\frac{4B_7}{1-\varepsilon}D_{n-1}\Bigg[\sum_{m=2}^{s}\rho^s\varepsilon_{m+1}(\varepsilon_{m+1}+\sum_{j>m+1}\varepsilon_j^2\sigma^{j-(m+1)})+$$
$$+\sum_{m=s+1}^{\infty}\rho^m\varepsilon_{m+1}(\varepsilon_{m+1}+\sum_{j>m+1}\varepsilon_j^2\sigma^{j-(m+1)})\Bigg]\leq$$

$$\leq \frac{4B_7(\sqrt{\rho})^s}{1-\varepsilon} D_{n-1} \left[\sum_{m=2}^{\infty} (\sqrt{\rho})^m \varepsilon_{m+1}^2 + \sum_{j=3}^{\infty} \varepsilon_j^2 \sum_{m=2}^{j-1} (\sqrt{\rho})^{m+1} \varepsilon_{m+1} \sigma^{j-m-1} \right]$$

$$\leq B_8 D_{n-1} (\sqrt{\rho})^s \left(\sum_{j=2}^{\infty} \varepsilon_j^2 \sigma^{j-1} \right),$$

where in the last step we used that $\sigma = \rho^{1/4} > \rho$, and hence

$$\sum_{m=2}^{j-1} (\sqrt{\rho})^{m+1} \varepsilon_{m+1} \sigma^{j-m-1} \leq \sigma^j \sum_m \left(\frac{\sqrt{\rho}}{\sigma} \right)^{m+1} \leq \frac{1}{1-\sqrt{\rho}/\sigma} \sigma^j.$$

We obtain the same estimates for the sum $\sum_{I \subset ([0,1/2] \cap J)}$ by symmetry. Thus, in view of (5.32)–(5.35) we have

$$D_n \leq \frac{16}{\pi(1-\varepsilon)^2} + 2B_8 D_{n-1} (\sqrt{\rho})^s.$$

Now if $s \geq M = M(\varepsilon) := 4 + 2\log(4B_8)/\log(1/\rho)$, then this gives

$$D_n \leq \frac{16}{\pi(1-\varepsilon)^2} + \frac{D_{n-1}}{2}, \tag{5.37}$$

while for $s \leq M$ we have

$$\varepsilon_1 \geq t_s \geq \left(\frac{1-\varepsilon}{2} \right)^M,$$

in which case (5.23) is clearly true with

$$D = \left(\frac{2}{1-\varepsilon} \right)^M.$$

Using induction we can infer from (5.37) that for all n

$$D_n \leq \left(\frac{2}{1-\varepsilon} \right)^M,$$

and the proof is over.

∎

After this we can turn to the proof of Theorem 5.1.

Proof of Theorem 5.1

The necessity of the condition follows from Corollary 3.3. Indeed, if $\sup_j \varepsilon_j = 1$, then with $t_k = 2^{-k}$ we can apply this corollary, and since $[t_k, t_{k-1}] \setminus E$ contains then for infinitely many k the whole interval $[t_k, t_{k-1}]$ (i.e. in Corollary 3.3 we have $\gamma_k^* = 1$ for infinitely many k), it follows that in this case the Green function does not satisfy the Lip 1/2 condition in (2.9). Thus, if the Green function $g_{\mathbf{C} \setminus C}$ is in Lip 1/2, then our basic assumption that there is an $\varepsilon < 1$ with $\varepsilon_j \leq \varepsilon$ is satisfied. Now consider the numbers t_n from (5.3). Since an interval of length $\varepsilon_{k+1} t_k$ is missing from $\mathcal{C}$ in the interval $[t_{k+1}, t_k]$, and the numbers t_k satisfy the conditions $(1-\varepsilon)t_k/2 \le t_{k+1} \le t_k/2$, we can invoke again Corollary 3.3 to conclude that the condition $\sum_j \varepsilon_j^2 < \infty$ is necessary for the Green function to belong to Lip 1/2.

To prove the sufficiency of $\sum_j \varepsilon_j^2 < \infty$ it is enough to prove that

under this assumption for the equilibrium measure we have $\mu_{\mathcal{C}}(I) \le C\sqrt{|I|}$ for any interval I

(see the proof of Theorem 2.2, in particular the argument between (2.57) and (2.59)).

This we shall prove in two parts: in Part I we verify this statement for intervals of the form $I = [0, \delta]$, i.e. Part I verifies the sufficiency part of Theorem 5.3. In Part II we give the proof for general I's by reducing the problem to the one discussed in Part I.

Proof of the sufficiency in Theorem 5.1, Part II.

Let η be a positive number, and consider all sequences $\{\varepsilon_j\}$ for which $\sum_j \varepsilon_j^2 \le \eta$ is satisfied. Let $D_n = D_n(\eta)$ be the smallest constant with the property, that for all such sequences $\{\varepsilon_j\}$ we have for any $\delta \le 1/2$ the inequality

$$\mu_{\mathcal{C}_n}([0,\delta]) \le D_n\sqrt{\delta}, \tag{5.38}$$

where $\mathcal{C}_n = \mathcal{C}_n(\varepsilon_1, \varepsilon_2, \ldots)$ is the set at the n-th level of the Cantor construction for the sequence $\{\varepsilon_j\}$. Clearly, since $\mathcal{C}_0 = [0,1]$,

$$d\mu_{\mathcal{C}_0}(t) = \frac{1}{\pi\sqrt{t(1-t)}}dt,$$

we have $D_0 \le 1$. We claim that if η is sufficiently small, then the numbers $D_n(\eta)$ form a bounded sequence, and in the limit $n \to \infty$ we obtain

$$\mu_{\mathcal{C}}([0,\delta]) \le D\sqrt{\delta} \tag{5.39}$$

provided $\sum_j \varepsilon_j^2 \le \eta$.

Thus, our aim is to prove the boundedness of $\{D_n(\eta)\}_n$ for sufficiently small η. To this end we shall estimate D_n in terms of D_{n-1}.

Note that if $\sum_j \varepsilon_j^2 \le \eta \le 1/4$, then each number ε_j is at most $1/2$. Thus we can see that our basic assumption $\varepsilon_j \le \varepsilon$ is satisfied with $\varepsilon = 1/2$. For the sake of easier reading in the discussion below we shall only use that $\varepsilon_j \le \varepsilon$ for some $\varepsilon < 1$, although we could write everywhere $\varepsilon = 1/2$.

The equilibrium measure $\mu_{\mathcal{C}_n}$ is obtained from $\mu := \mu_{[0,1]}$ by taking balayage onto $\mathcal{C}_n$. Let us denote the balayage of a measure ν onto $\mathcal{C}_n$ by $\overline{\nu}$. Thus, $\mu_{\mathcal{C}_n} = \overline{\mu}$. Let M be the largest number with the property $2\delta \le t_{M+3}$, where the numbers t_n have been defined in (5.3). If there is no such M then $\delta > t_3/2 \ge (1-\varepsilon)^3/16$, in which case (5.38) is true with $D_n = \sqrt{16/(1-\varepsilon)^3}$, so without loss of generality we may assume the existence of M. With this M we write

$$\overline{\mu} = \mu|_{\mathcal{C}_n} + \overline{\mu|_{[0,t_{M+1}]\setminus\mathcal{C}_n}} + \sum_{k=0}^{M} \overline{\mu|_{[t_{k+1},t_k]\setminus\mathcal{C}_n}}. \tag{5.40}$$

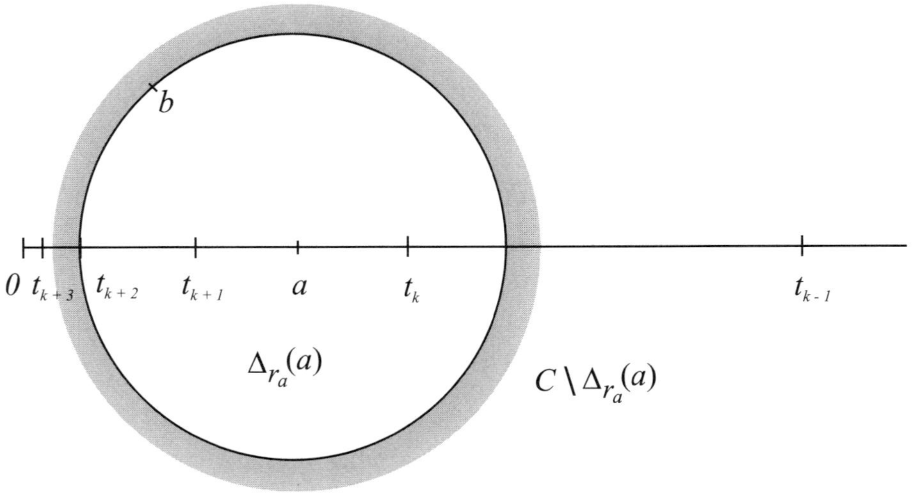

Figure 12:

Here, since $d\mu(t) = dt/\pi\sqrt{t(1-t)}$, we have

$$\overline{\mu\big|[t_{k+1}, t_k] \setminus \mathcal{C}_n}([0,\delta]) = \int_{[t_{k+1},t_k]\setminus \mathcal{C}_n} \frac{1}{\pi\sqrt{a(1-a)}} \overline{\delta_a}([0,\delta])da. \tag{5.41}$$

For $a \in [t_{k+1}, t_k] \setminus \mathcal{C}_n$ let $r_a = a - t_{k+2}$, for which we have the inequalities $t_{k+2} \leq r_a \leq t_k$, which imply

$$\frac{(1-\varepsilon)^2}{4} \leq \frac{r_a}{t_k} \leq 1. \tag{5.42}$$

The balayage $\overline{\delta_a}$ can be formed in two steps: first take the balayage of δ_a onto $(\mathbf{C} \setminus \Delta_{r_a}(a)) \cup \mathcal{C}_n$ — let this measure be ν_a —, and then take the balayage of ν_a onto $\mathcal{C}_n$. The measure ν_a is supported on $(\mathcal{C}_n \cap \Delta_{r_a}(a)) \cup C_{r_a}(a)$ (see Figure 12), and in calculating $\overline{\delta_a}([0,\delta])$, in the second step we only have to take the balayage onto $\mathcal{C}_n$ of that part of ν_a that lies on $C_{r_a}(a)$.

For fixed $0 \leq k \leq M$ let $\widehat{\nu}$ be the balayage of a measure ν onto the set $[0, t_{k+3}] \cap \mathcal{C}_n$. Then $\overline{\nu}([0,\delta]) \leq \widehat{\nu}([0,\delta])$, therefore, for $a \in [t_{k+1}, t_k] \setminus \mathcal{C}_n$ we have

$$\overline{\delta_a}([0,\delta]) = \overline{\nu_a\big|C_{r_a}(a)}([0,\delta]) \leq \int_{C_{r_a}(a)} \widehat{\delta_b}([0,\delta])d\nu_a(b). \tag{5.43}$$

However, for $b \in C_{r_a}(a)$ we have $\mathrm{dist}(b, [0, t_{k+3}]) \geq t_{k+3}$ (see Figure 12), therefore it follows from Harnack's inequality (2.31) that

$$\frac{1}{3}\widehat{\delta_\infty} \leq \widehat{\delta_b} \leq 3\widehat{\delta_\infty}.$$

But $\widehat{\delta_\infty} = \mu_{[0,t_{k+3}]\cap\mathcal{C}_n}$, the equilibrium measure of $[0, t_{k+3}] \cap \mathcal{C}_n$, and this latter set is nothing else than

$$\mathcal{C}_{n-k-3}([0, t_{k+3}]; \varepsilon_{k+3+1}, \varepsilon_{k+3+2}, \ldots) = t_{k+3}\mathcal{C}_{n-k-3}(\varepsilon_{k+3+1}, \varepsilon_{k+3+2}, \ldots).$$

Thus, by the definition of the numbers D_i we have

$$\widehat{\delta_b}([0,\delta]) \leq 3\mu_{[0,t_{k+3}]\cap\mathcal{C}_n}([0,\delta]) \leq 3D_{n-k-3}\sqrt{\frac{\delta}{t_{k+3}}} \leq 3D_{n-1}\sqrt{\frac{\delta}{t_{k+3}}}. \tag{5.44}$$

All in all, we get from (5.43) – (5.44) for $a \in [t_{k+1}, t_k] \setminus \mathcal{C}_n$

$$\overline{\delta_a}([0,\delta]) \leq 3D_{n-1}\sqrt{\frac{\delta}{t_{k+3}}}\nu_a(C_{r_a}(a)). \tag{5.45}$$

The set $[t_{k+1}, t_k] \setminus \mathcal{C}_n$ can be written as

$$\bigcup_{m=0}^{n-k-1} \left([t_{k+1}, t_k] \cap (\mathcal{C}_{k+m} \setminus \mathcal{C}_{k+m+1})\right). \tag{5.46}$$

Let $a \in [t_{k+1}, t_k] \cap (\mathcal{C}_{k+m} \setminus \mathcal{C}_{k+m+1})$. We have

$$\begin{aligned} &\omega(a, C_{r_a}(a), \Delta_{r_a}(a) \setminus \mathcal{C}_n) = \omega\Big(a/t_k, C_{r_a/t_k}(a/t_k), \Delta_{r_a/t_k}(a/t_k) \setminus (\mathcal{C}_n/t_k)\Big) \\ \leq\ & \omega\Big(a/t_k, C_{r_a/t_k}(a/t_k), \Delta_{r_a/t_k}(a/t_k) \setminus (\mathcal{C}_n \cap [0,t_k])/t_k)\Big), \end{aligned}$$

and here the set $(\mathcal{C}_n \cap [0, t_k])/t_k$ is exactly $\mathcal{C}_{n-k}(\varepsilon_{k+1}, \varepsilon_{k+2}, \ldots)$, and the point a/t_k belongs to

$$\mathcal{C}_m(\varepsilon_{k+1}, \varepsilon_{k+2}, \ldots) \setminus \mathcal{C}_{m+1}(\varepsilon_{k+1}, \varepsilon_{k+2}, \ldots).$$

Since for r_a/t_k we have (5.42), we can apply Theorem 5.8, (5.15) and we obtain that with some $\rho < 1$

$$\omega(a, C_{r_a}(a), \Delta_{r_a}(a) \setminus \mathcal{C}_n) \leq 3B\rho^m \omega(1/2, C_0, \Delta_0 \setminus \mathcal{C}(\varepsilon_{k+m+1}, \varepsilon_{k+m+2}, \ldots)).$$

On applying Theorem 5.9 to the last factor we can conclude

$$\begin{aligned} \nu_a(C_{r_a}(a)) &= \omega(a, C_{r_a}(a), \Delta_{r_a}(a) \setminus \mathcal{C}_n) \leq \\ &\leq B_1\sigma^m(\varepsilon_{k+m+1} + \sum_{j>k+m+1} \varepsilon_j^2 \sigma^{j-(k+m+1)}), \end{aligned} \tag{5.47}$$

where $\sigma < 1$ and B_1 and σ depend only on the bound ε for the numbers ε_j.

In view of (5.41)–(5.47) we have for

$$\overline{\mu\big|[t_{k+1}, t_k] \setminus \mathcal{C}_n}([0,\delta]) \tag{5.48}$$

the bound

$$\begin{aligned} 3B_1 D_{n-1}\sqrt{\frac{\delta}{t_{k+3}}} \qquad & \sum_{m=0}^{n-k-1} \sigma^m(\varepsilon_{k+m+1} + \sum_{j>k+m+1} \varepsilon_j^2 \sigma^{j-(k+m+1)}) \times \\ & \times \int_{[t_{k+1},t_k]\cap(\mathcal{C}_{k+m}\setminus\mathcal{C}_{k+m+1})} \frac{1}{\pi\sqrt{a(1-a)}}da. \end{aligned}$$

In this last integral the set of integration consists of equal intervals of length $\varepsilon_{k+m+1}t_{k+m}$, the number of which is 1 if $m = 0$ and 2^{m-1} if $m \geq 1$. Thus, if $k \geq 1$, then the integral is bounded by

$$2^m \varepsilon_{k+m+1} t_{k+m} \frac{1}{\sqrt{t_{k+1}}} \leq \sqrt{\frac{2}{1-\varepsilon}} \varepsilon_{k+m+1}\sqrt{t_k}.$$

If, however, $k = 0$, then we also have to use that the l-th of these intervals from the right is contained in $[1/2, 1 - lt_{m+1}]$. Therefore in this case we can bound the integral as (use also $t_{m+1} \geq t_m(1-\varepsilon)/2$))

$$\varepsilon_{m+1} t_m \sum_{l=1}^{2^m} \frac{1}{\sqrt{lt_m(1-\varepsilon)/2}} \leq \sqrt{\frac{2}{1-\varepsilon}} \varepsilon_{m+1} 2\sqrt{2^m}\sqrt{t_m} \leq 2\sqrt{\frac{2}{1-\varepsilon}} \varepsilon_{m+1}.$$

All these yield that (5.48) is

$$\begin{aligned} &\leq 3B_1 D_{n-1} \sqrt{\frac{\delta}{t_{k+3}}} \sum_{m=0}^{n-k-1} \sigma^m \big(\varepsilon_{k+m+1} + \sum_{j>k+m+1} \varepsilon_j^2 \sigma^{j-(k+m+1)}\big) \times \\ &\qquad \times 2\sqrt{\frac{2}{1-\varepsilon}} \varepsilon_{k+m+1} \sqrt{t_k} \leq \\ &\leq \frac{24 B_1 D_{n-1}}{(1-\varepsilon)^2} \sqrt{\delta} \sum_{m=0}^{n-k-1} \sigma^m \varepsilon_{k+m+1} \big(\varepsilon_{k+m+1} + \sum_{j>k+m+1} \varepsilon_j^2 \sigma^{j-(k+m+1)}\big). \end{aligned}$$

Summing these for $k = 0, \ldots, M$ we obtain for

$$\overline{\mu\big|_{[t_{M+1}, 1] \setminus \mathcal{C}_n}}([0,\delta])$$

the estimate

$$\leq \frac{24 B_1 D_{n-1}}{(1-\varepsilon)^2} \sqrt{\delta} \sum_{k=0}^{M} \sum_{m=0}^{n-k-1} \sigma^m \varepsilon_{k+m+1} \big(\varepsilon_{k+m+1} + \sum_{j>k+m+1} \varepsilon_j^2 \sigma^{j-(k+m+1)}\big).$$

The last double sum is at most

$$\sum_{j=1}^{\infty} \varepsilon_j^2 \big(\sum_m \sigma^m + \sum_l l\sigma^l\big) \leq \frac{2}{(1-\sigma)^2} \sum_{j=1}^{\infty} \varepsilon_j^2.$$

If we calculate $\overline{\mu}([0,\delta])$ on the right of (5.40), the sum of the first two terms is at most

$$\begin{aligned} \mu(\mathcal{C}_n \cap [0,\delta]) \;&+\; \overline{\mu\big|_{[0, t_{M+1}] \setminus \mathcal{C}_n}}([0,\delta]) \leq \mu([0, t_{M+1}]) \leq \sqrt{t_{M+1}} \leq \\ &\leq \sqrt{\left(\frac{2}{1-\varepsilon}\right)^3 t_{M+4}} \leq \sqrt{\left(\frac{2}{1-\varepsilon}\right)^3 2\delta} \leq \frac{4}{(1-\varepsilon)^2}\sqrt{\delta}. \end{aligned}$$

Thus,

$$\mu_{\mathcal{C}_n}([0,\delta]) \leq \sqrt{\delta} \left(\frac{4}{(1-\varepsilon)^2} + \frac{24 B_1 D_{n-1}}{(1-\varepsilon)^2} \frac{2}{(1-\sigma)^2} \sum_j \varepsilon_j^2 \right), \tag{5.49}$$

and this estimates is correct regardless if the number M with property $t_{M+4} < 2\delta \leq t_{M+3}$ exists or not (in which case, as we have seen, $\delta \geq t_3/2 = (1-\varepsilon)^3/16$ and hence $\sqrt{\delta}(4/(1-\varepsilon)^2) \geq 1$).

Here the numbers B_1 and $\sigma < 1$ depend only on the bound ε of the numbers ε_j. Recall, that if $\sum_j \varepsilon_j^2 \leq \eta$ and $\eta \leq 1/4$, then we can set everywhere in this proof $\varepsilon = 1/2$, and hence B_1 and $\sigma < 1$ are some absolute numbers. Thus, for $\sum_j \varepsilon_j^2 \leq \eta \leq 1/4$ we obtain

$$D_n(\eta) \leq 16 + \frac{192 B_1 D_{n-1}}{(1-\sigma)^2}\eta,$$

and so if $\eta \leq 1/4$ also satisfies

$$\eta \leq \frac{(1-\sigma)^2}{4 \cdot 192 B_1}, \tag{5.50}$$

then we get

$$D_n(\eta) \leq 16 + \frac{D_{n-1}(\eta)}{2},$$

from which it follows by induction that

$$D_n(\eta) \leq 32. \tag{5.51}$$

This estimate, and with it the inequality (5.39), was obtained under the assumption $\eta \leq q$ where q is the number on the right of (5.50), and therefore the proof of Theorem 5.1 is complete provided the ε_j's satisfy $\sum_j \varepsilon_j^2 \leq q$.

Let now $\{\varepsilon_j\}$ be an arbitrary sequence from the interval $[0, 1)$ with $\sum_j \varepsilon_j^2 < \infty$. Then there is an $\varepsilon < 1$ such that for all i we have $\varepsilon_j \leq \varepsilon$. Choose the smallest N such that $\sum_{j>N} \varepsilon_j^2 < q$ is satisfied, where q is the number on the right of (5.50). If

$$\delta \geq \frac{1}{2}\left(\frac{1-\varepsilon}{2}\right)^N, \tag{5.52}$$

then

$$\mu_C([0,\delta]) \leq 1 \leq \sqrt{\frac{2 \cdot 2^N}{(1-\varepsilon)^N}}\sqrt{\delta}.$$

If, however, (5.52) is not true, then $\delta < t_N/2$, and so we obtain from (5.51)

$$\begin{aligned}\mu_C([0,\delta]) &\leq \mu_{[0,t_N]\cap C}([0,\delta]) = \mu_{C(\varepsilon_{N+1},\varepsilon_{N+2},\ldots)}([0,\delta/t_N]) \\ &\leq 32\sqrt{\frac{\delta}{t_N}} \leq 32\sqrt{\frac{2^N}{(1-\varepsilon)^N}}\sqrt{\delta},\end{aligned}$$

and these show that (5.39) is true with $D = 32 \cdot 2^{N/2}/(1-\varepsilon)^{N/2}$.

■

Proof of the sufficiency in Theorem 5.1, Part II.
In Part I of the proof we verified that if $\sum_i \varepsilon_i^2 < \infty$ and $I = [0, \delta]$, then

$$\mu_{C(\varepsilon_1,\varepsilon_2,\ldots)}(I) \leq C\sqrt{|I|}. \tag{5.53}$$

In this second part we prove the same inequality but for arbitrary $I \subseteq [0, 1]$. Roughly speaking, we shall show that if the length of I is fixed, then the largest value of $\mu_C(I) = \mu_{C(\varepsilon_1,\varepsilon_2,\ldots)}(I)$ occurs at the endpoint 0, i.e. when $I = [0, |I|]$, and hence (5.53) will prove the general case. To carry out this program, we shall show – very roughly speaking – that the equilibrium measure μ_C behaves as if it was given by a density function that decreases on $[0,1/2]$ and increases on $[1/2,1]$. This is of course not true, but if we measure the mass of the equilibrium measure on the individual subintervals lying on the n-th level of the Cantor construction, then this statement is correct to a certain extent.

First we shall need to get some information on taking balayage measures onto symmetric sets.

Let $\mathrm{Bal}\Big(\rho,F\Big)$ denote the balayage of the measure ρ onto the set F (i.e. out of $\mathbf{C}\setminus F$). For an interval $(\alpha,\beta)\subset(0,\infty)$ we shall also need to take the balayage of a measure ρ onto the unbounded set $\mathbf{R}\setminus\big((\alpha,\beta)\cup(-\beta,-\alpha)\big)$, which we define as the limit

$$\mathrm{Bal}\Big(\rho,\mathbf{R}\setminus\big((\alpha,\beta)\cup(-\beta,-\alpha)\big)\Big)=\lim_{R\to\infty}\mathrm{Bal}\Big(\rho,[-R,R]\setminus\big((\alpha,\beta)\cup(-\beta,-\alpha)\big)\Big).$$

According to formula (2.25), if $a\in(\alpha,\beta)$ then $\mathrm{Bal}\Big(\delta_a,[-R,R]\setminus\big((\alpha,\beta)\cup(-\beta,-\alpha)\big)\Big)$ is given by the density

$$h_{R,a}(t)=\frac{\sqrt{|a^2-\alpha^2||a^2-\beta^2||a^2-R^2|}}{\pi\sqrt{|t^2-\alpha^2||t^2-\beta^2||t^2-R^2|}}\frac{|t-\zeta||t-\xi|}{|a-\zeta||a-\xi|}\frac{1}{|t-a|},$$

where $\zeta=\zeta_{R,a}\in(-\beta,-\alpha)$ and $\xi=\xi_{R,a}\in(-\infty,-R)\cup(R,\infty)$, and among others these numbers have the property that

$$\int_{-\beta}^{-\alpha}h_{R,a}(t)\mathrm{sign}(t-\zeta_{R,a})dt=0.$$

Here $|a^2-R^2|/|t^2-R^2|$ uniformly tends to 1 on compact subsets of the real line as $R\to\infty$, and elementary estimates show that for $|t|\in[2\beta,R/2]$

$$h_{R,a}(t)\le\frac{C}{t^2}$$

while for $|t|\in[R/2,R]$

$$h_{R,a}(t)\le\frac{C}{R\sqrt{R^2-t^2}}$$

with a constant C which is independent of $a\in(\alpha,\beta)$ and $R>2\beta$. These show that the measures $h_{R,a}(t)dt$ are uniformly "small" outside some large intervals, and these properties easily imply that on compact subsets of the real line

$$\lim_{R\to\infty}h_{R,a}(t)=h_a(t):=\frac{\sqrt{|a^2-\alpha^2||a^2-\beta^2|}}{\pi\sqrt{|t^2-\alpha^2||t^2-\beta^2|}}\frac{|t-\zeta_a|}{|a-\zeta_a|}\frac{1}{|t-a|},\tag{5.54}$$

where $\zeta_a\in[-\beta,-\alpha]$ satisfies

$$\int_{-\beta}^{-\alpha}h_a(t)\mathrm{sign}(t-\zeta_a)dt=0.$$

Note that there is a unique ζ_a with this property and so $\zeta_{R,a}$ must tend to ζ_a as $R\to\infty$. It also follows that

$$h_a(t)\le\frac{C}{t^2},\qquad t\ge 2\beta\tag{5.55}$$

with some C independent of $a \in (\alpha, \beta)$,

$$\int_{\mathbf{R}\setminus\big((\alpha,\beta)\cup(-\beta,-\alpha)\big)} h_a(t)dt = \lim_{R\to\infty} \int_{[-R,R]\setminus\big((\alpha,\beta)\cup(-\beta,-\alpha)\big)} h_{R,a}(t)dt = 1$$

and

$$\int_{\mathbf{R}\setminus\big((\alpha,\beta)\cup(-\beta,-\alpha)\big)} h_a(t)\log\frac{1}{|x-t|}dt$$
$$= \lim_{R\to\infty} \int_{[-R,R]\setminus((-\beta,-\alpha)\cup(\alpha,\beta))} h_{R,a}(t)\log\frac{1}{|x-t|}dt$$

uniformly for x lying in any compact subset of the real line. Now recall that for $z \in [-R,R]\setminus\big((\alpha,\beta)\cup(-\beta,-\alpha)\big)$ the last integral is equal to

$$\log\frac{1}{|x-a|} + c_R$$

with some constant c_R (actually $c_R = g_{\mathbf{C}\setminus([-R,R]\setminus((-\beta,-\alpha)\cup(\alpha,\beta)))}(a)$), hence we can conclude that the numbers c_R have a limit c as $R \to \infty$, and that for any $x \in \mathbf{R}\setminus((-\beta,-\alpha)\cup(\alpha,\beta))$

$$\int_{\mathbf{R}\setminus\big((\alpha,\beta)\cup(-\beta,-\alpha)\big)} h_a(t)\log\frac{1}{|x-t|}dt = \log\frac{1}{|x-a|} + c.$$

If ρ is an arbitrary measure on $[0,\infty)$, then we form the balayage measure $\mathrm{Bal}\Big(\rho, \mathbf{R}\setminus\big((\alpha,\beta)\cup(-\beta,-\alpha)\big)\Big)$ by adding to $\rho|_{\mathbf{R}\setminus((\alpha,\beta)\cup(-\beta,-\alpha))}$ the balayage of $\rho|_{(\alpha,\beta)}$ onto $\mathbf{R}\setminus\big((\alpha,\beta)\cup(b,-\alpha)\big)$, and this latter measure is the same as

$$\int_\alpha^\beta \mathrm{Bal}\Big(\delta_a, \mathbf{R}\setminus\big((\alpha,\beta)\cup(b,-\alpha)\big)\Big)d\rho(a), \tag{5.56}$$

hence it is given by the density

$$\int_\alpha^\beta h_a(t)d\rho(a). \tag{5.57}$$

Thus, what we have said above implies that

$$\Big\|\mathrm{Bal}\Big(\rho, \mathbf{R}\setminus\big((\alpha,\beta)\cup(-\beta,-\alpha)\big)\Big)\Big\| = \|\rho\|$$

($\|\rho\|$ denoting the total mass of ρ), and for $x \in R\setminus\big((\alpha,\beta)\cup(-\beta,-\alpha)\big)$

$$\int \log\frac{1}{|x-t|}d\mathrm{Bal}\Big(\rho, \mathbf{R}\setminus\big((\alpha,\beta)\cup(-\beta,-\alpha)\big)\Big)(t) = \int \log\frac{1}{|x-t|}d\rho(t) + c_\rho \tag{5.58}$$

with some constant c_ρ.

Let us say that a measure ν on the real line is *right-dominant* if for any interval $A \subset (0,\infty)$ we have $\nu(-A) \le \nu(A)$. The form of h_a in (5.54) shows that $h_a(-t) \le h_a(t)$ for $t \ge 0$. Therefore the measure $h_a(t)dt$ is right-dominant for any $a \in (\alpha,\beta)$, and hence so is any measure of the form (5.56) – (5.57). Thus, we can conclude that if ρ is a measure on $(0,\infty)$, then $\mathrm{Bal}\Big(\rho, \mathbf{R}\setminus\big((\alpha,\beta)\cup(-\beta,-\alpha)\big)\Big)$ is right- dominant. But if ν is any right-dominant measure then we can write it as $\nu = \nu_{\mathrm{symm}} + \rho$ where ν_{symm} is a measure symmetric onto the origin and ρ is a measure on $(0,\infty)$. Since $\mathrm{Bal}\Big(\nu_{\mathrm{symm}}, \mathbf{R}\setminus\big((\alpha,\beta)\cup(-\beta,-\alpha)\big)\Big)$ is clearly symmetric onto the origin, we can conclude that $\mathrm{Bal}\Big(\nu, \mathbf{R}\setminus\big((\alpha,\beta)\cup(-\beta,-\alpha)\big)\Big)$ is right-dominant.

Thus, we have proved that

if ν is right-dominant and we take the balayage of ν onto $\mathbf{R}\setminus\big((\alpha,\beta)\cup(-\beta,-\alpha)\big)$, then the balayage measure is also right-dominant.

Next we want to extend this to taking balayage measures for more general sets. Let E be compact, symmetric with respect to the origin and consisting of finitely many intervals. We want to show that

if ν is a right-dominant measure on the real line then its balayage onto E is also right-dominant.

This will follow from what we have just established if we can show that the balayage onto E can be obtained by repeatedly taking balayage onto some sets of the form $\mathbf{R}\setminus\big((\alpha,\beta)\cup(-\beta,-\alpha)\big)$. More precisely let

$$(\mathbf{R}\setminus\{0\})\setminus E = \bigcup_{i=1}^{N}\big((\alpha_i,\beta_i)\cup(-\beta_i,-\alpha_i)\big),$$

and let $I_0, I_1, \ldots$ be the sequence of intervals that we obtain by repeating the finite sequence $(\alpha_1,\beta_1),\ldots,(\alpha_N,\beta_N)$ infinitely many times. We claim that if $\nu_0 = \nu$ and

$$\nu_{n+1} = \mathrm{Bal}\Big(\nu_n, \mathbf{R}\setminus(I_n\cup(-I_n))\Big), \qquad n = 0,1,\ldots$$

then $\nu_n \to \mathrm{Bal}\Big(\nu, E\Big)$ in the weak* topology. This will already prove that if ν is right-dominant then $\mathrm{Bal}\Big(\nu, E\Big)$ is also right-dominant, for we have seen above that this property is inherited from ν_n to ν_{n+1}, and it is also clear that the weak* limit of right-dominant measures is right-dominant (recall also that $\mathrm{Bal}\Big(\nu, E\Big)$ is absolutely continuous with respect to linear Lebesgue measure).

It is clear that $\nu_{n+1}\big|_E \ge \nu_n\big|_E$, and first we show that for any I_j we have $\nu_n(I_j)\to 0$, $\nu_n(-I_j)\to 0$ as $n\to\infty$. Indeed, if this was not the case, then, say, there was an infinite sequence n_k of the natural numbers and an $\eta>0$ such that $\nu_{n_k}(I_j)\ge\eta$ for all k. Since there are two subintervals of E attached to I_j if I_j is finite and one subinterval of E attached to I_j if I_j is an infinite interval, it easily follows that there is a $\delta_j>0$ such that if ρ is any measure on I_j, then by forming the balayage of ρ onto $\mathbf{R}\setminus(I_j\cup(-I_j))$, at least $\delta_j\rho(I_j)$ of the balayage mass will lie on these attached subintervals, i.e.

$$\mathrm{Bal}\Big(\rho, \mathbf{R}\setminus(I_j\cup(-I_j))\Big)(E) \ge \delta_j\rho(I_j).$$

But this implies that

$$\nu_{n_k+1}(E) \geq \nu_{n_k}(E) + \delta_j \nu_{n_k}(I_j) \geq \nu_{n_k}(E) + \delta_j \eta$$

for all $k = 1, 2, \ldots$, which is impossible for the measures ν_n all have the same finite total mass.

Thus, we get $\nu_n((\alpha_j, \beta_j)) \to 0$ and $\nu_n((-\beta_j, -\alpha_j)) \to 0$ for all $j = 1, \ldots, N$, and then (5.55) shows that if β_N is the largest of the β_j's, then

$$\frac{d\nu_n(t)}{dt} \leq \frac{\varepsilon_n}{t^2}, \qquad t > 2\beta_N$$

with some sequence $\varepsilon_n \to 0$. Now these easily imply that for the measure

$$\widehat{\nu} := \lim_{n\to\infty} \nu_n\big|_E$$

(recall that the measures $\nu_n\big|_E$ are increasing so this limit exists) we have $\widehat{\nu}(E) = \|\nu\|$, $\nu(\mathbf{R} \setminus E) = 0$, and for $x \in \mathrm{Int}(E)$

$$\int \log \frac{1}{|x-t|} d\widehat{\nu}(t) = \lim_{n\to\infty} \int \log \frac{1}{|x-t|} d\nu_n(t). \tag{5.59}$$

We have seen in (5.58) that here

$$\int \log \frac{1}{|x-t|} d\nu_n(t) = \int \log \frac{1}{|x-t|} d\nu_{n-1}(t) + c_{\nu_{n-1}},$$

and so induction shows that

$$\int \log \frac{1}{|x-t|} d\nu_n(t) = \int \log \frac{1}{|x-t|} d\nu(t) + c_n$$

with some constants c_n. But then we can conclude from (5.59) that the sequence $\{c_n\}$ converges to some c as $n \to \infty$ and for $x \in \mathrm{Int}(E)$

$$\int \log \frac{1}{|x-t|} d\widehat{\nu}(t) = \int \log \frac{1}{|x-t|} d\nu(t) + c.$$

Since $\|\widehat{\nu}\| = \|\nu\|$ is also true, this shows that $\widehat{\nu}$ is the balayage of ν onto E as we have claimed.

So far we have shown that if E is symmetric and consists of finitely many (finite) intervals, then the balayage of a right-dominant measure onto E is again right-dominant.

After this let us turn to Cantor sets $\mathcal{C} = \mathcal{C}(\varepsilon_1, \varepsilon_2, \ldots)$. In analogy with right-dominance let us say that a measure ν is right(a)-dominant if for any interval $A \subset (a, \infty)$ we have $\nu(a - A) \leq \nu(A)$, which is of course the same as the translated measure $\nu(\cdot - a)$ being right-dominant. In a similar manner let us talk about left(a)-dominant measures. We claim that if $\mu_{\mathcal{C}}$ is the equilibrium measure of the Cantor set $\mathcal{C} = \mathcal{C}(\varepsilon_1, \varepsilon_2, \ldots)$, then for any k the measure $\mu_{\mathcal{C}}\big|_{[0,\,t_k]}$ is left($t_k/2$)-dominant (see (5.3) for the definition of the numbers t_n). In fact, if we take the balayage of $\mu_{\mathcal{C}}$ onto $\mathcal{C} \cap [0, t_k]$ then we obtain the equilibrium measure of the latter set, i.e.

$$\mu_{\mathcal{C}\cap[0,t_k]} = \mu_{\mathcal{C}}\big|_{[0,\,t_k]} + \mathrm{Bal}\Big(\mu_{\mathcal{C}}\big|_{(t_k,\,\infty)}, \mathcal{C} \cap [0, t_k]\Big). \tag{5.60}$$

Since $\mathcal{C} \cap [0, t_k]$ is symmetric onto $t_k/2$, and a-priory the measure $\mu_{\mathcal{C}}\Big|_{(t_k,\infty)}$ is right$(t_k/2)$-dominant, we get that the balayage measure on the right of (5.60) is right$(t_k/2)$-dominant. Using the symmetry of $\mu_{\mathcal{C}\cap[0,t_k]}$ we get from here the left$(t_k/2)$-dominance of $\mu_{\mathcal{C}}\Big|_{[0,\,t_k]}$, as we claimed.

Now let I be any of the 2^n intervals of length t_n that we obtain at the n-th level of the Cantor construction. We want to estimate $\mu_{\mathcal{C}}(I)$. By symmetry, we may assume that $I \subset [0, 1/2]$. If k is the largest index with $I \subset [0, t_k]$ and $k < n$, then $I \subset [t_k/2, t_k]$, and hence by the left$(t_k/2)$-dominance of $\mu_{\mathcal{C}}$ we obtain that $\mu_{\mathcal{C}}(t_k/2 - I) \geq \mu_{\mathcal{C}}(I)$, and here $t_k/2 - I$ is another subinterval of length t_n at the n-th level of the Cantor construction. This means that if I is not the leftmost interval (i.e $[0, t_n]$) at the n-th level, then there is another interval to the left of I which carries at least as much mass from the equilibrium measure as I. Thus, for all I we must have $\mu_{\mathcal{C}}(I) \leq \mu_{\mathcal{C}}([0, t_n])$, and for the latter value we have proved in Part I of the proof the estimate $\leq C\sqrt{t_n}$, by which we have verified

$$\mu_{\mathcal{C}}(I) \leq C\sqrt{|I|}. \tag{5.61}$$

Finally, let $I \subseteq [0, 1]$ be an arbitrary interval. In estimating $\mu_{\mathcal{C}}(I)$ we may assume that the left endpoint of I is a right endpoint of one of the contiguous subintervals to $\mathcal{C}$ (if this is not the case, then either drop that part of I around this endpoint that does not belong to $\mathcal{C}$ or, in case when the left endpoint of I is in the set $\mathcal{C}$, add a tiny subinterval to $\mathcal{C}$), and similarly we may assume that the right endpoint of I coincides with a left endpoint of one of the contiguous subintervals to $\mathcal{C}$. Let $(a_k, b_k) \subset I$ be the contiguous subinterval to $\mathcal{C}$ which is omitted at the lowest possible level, say at the k-th step of the Cantor construction. Note that this is uniquely determined, for if $(c_k, d_k) \subset I$ was another subinterval omitted at the k-th step of the Cantor construction, then between (a_k, b_k) and (c_k, d_k) there would lie another contiguous subinterval omitted at the $(k-1)$-st step, and this is not possible by the choice of k. Now $[a_k - t_k, a_k]$ and $[b_k, b_k + t_k]$ are the two subintervals attached to (a_k, b_k) that lie on level k of the Cantor construction (see Figure 13), and we must have $I \subset [a_k - t_k, b_k + t_k]$. Let $I_1 = I \cap [a_k - t_k, a_k]$ and $I_2 = I \cap [b_k, b_k + t_k]$. Then $I \cap \mathcal{C} = (I_1 \cap \mathcal{C}) \cup (I_2 \cap \mathcal{C})$, and so it is sufficient to estimate separately the measures $\mu_{\mathcal{C}}(I_1)$ and $\mu_{\mathcal{C}}(I_2)$. Consider e.g. the latter one, and let n be the largest index with $I_2 \subseteq [b_k, b_k + t_n]$. Since $t_{n+1} = t_n(1 - \varepsilon_{n+1})/2 \geq t_n(1 - \varepsilon)/2$ where $\varepsilon = \sup_i \varepsilon_i < 1$, it follows that

$$t_n(1 - \varepsilon)/2 \leq |I_2| \leq t_n.$$

But $[b_k, b_k + t_n]$ is one of the subintervals at the n-th level of the Cantor construction, and for this we have already verified (5.61). Thus we can conclude

$$\mu_{\mathcal{C}}(I_2) \leq \mu_{\mathcal{C}}([b_k, b_k + t_n]) \leq C\sqrt{t_n} \leq C\sqrt{\frac{2}{1-\varepsilon}}\sqrt{|I_2|} \leq C\sqrt{\frac{2}{1-\varepsilon}}\sqrt{|I|}.$$

The estimate of $\mu_{\mathcal{C}}(I_1)$ is completely analogous, and these prove (5.61) for I.

■

Proof of Theorem 5.3

The sufficiency of the condition has been verified in the previous proof. The necessity follows in the same manner from the analogue of Corollary 3.3 (more precisely

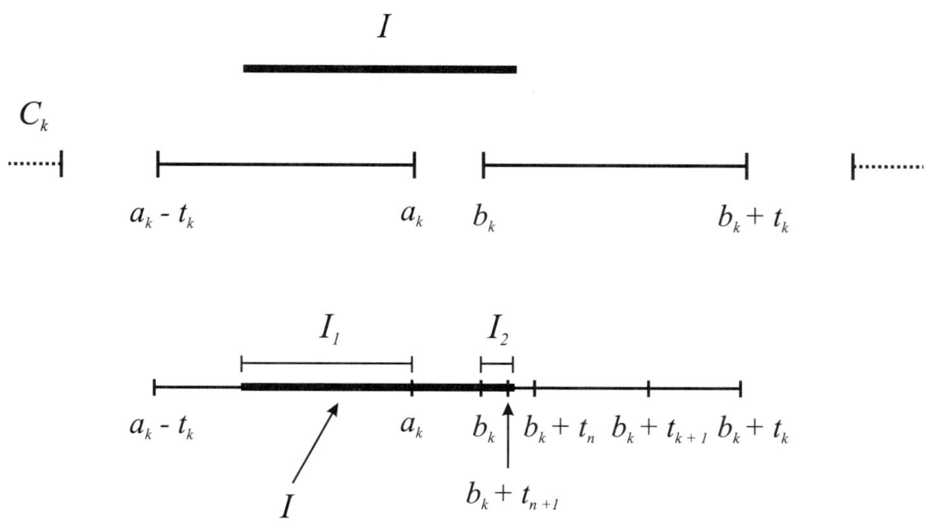

Figure 13:

from the statement given right after Corollary 3.3) as the necessity of Theorem 5.1 followed from Corollary 3.3.

■

For later use we need the following technical extension of Theorem 5.1.

Corollary 5.10 *Let $\sum_i \varepsilon_i^2 < \infty$. Then there is a constant C such that for all $s \geq 1$ we have*

$$g_{\mathbf{C}\setminus(\mathcal{C}(\varepsilon_1,\varepsilon_2,\ldots)\cap[t_s,1])}(y) \leq C\sqrt{|y|}, \qquad |y| \geq t_s, \tag{5.62}$$

where the numbers t_s are the ones from (5.3).

Note that we have here a Green function that is larger than the one in Theorem 5.1.

Proof of Corollary 5.10. Let $\mathcal{C}'(\varepsilon_1, \varepsilon_2, \ldots) = \mathcal{C}(\varepsilon_1, \varepsilon_2, \ldots) \setminus [0, t_s]$. Then the leftmost point in $\mathcal{C}'(\varepsilon_1, \varepsilon_2, \ldots)$ is t_s^*, and it is again enough to show that if $E = \mathcal{C}'(\varepsilon_1, \varepsilon_2, \ldots)$, then $\mu_E([t_s^*, t_s^* + \delta]) \leq C\sqrt{\delta}$ with a C that is independent of s. This follows from Theorem 5.3 if $\delta \geq t_{s+3}$. Indeed, μ_E is obtained from the equilibrium measure of $\mathcal{C}(\varepsilon_1, \varepsilon_2, \ldots)$ by taking the balayage of the latter one onto E, and in taking this balayage measure only the part that lies in $[0, t_s]$ is actually moving. Thus,

$$\begin{aligned}\mu_E([t_s^*, t_s^* + \delta]) &\leq \mu_{\mathcal{C}(\varepsilon_1,\varepsilon_2,\ldots)}([t_s^*, t_s^* + \delta]) + \mu_{\mathcal{C}(\varepsilon_1,\varepsilon_2,\ldots)}([0, t_s]) \\ &\leq C\sqrt{\delta} + C\sqrt{t_s} \leq C\sqrt{\delta}.\end{aligned}$$

Unfortunately such an easy argument does not work if $\delta < t_{s+3}$, and in that case we have to follow the proof of Theorem 5.1; which works for $\delta < t_{s+3}$ because then the proof just used local portions of the set $\mathcal{C}$, and they are exactly the same for $\mathcal{C}'$. We shortly indicate the necessary changes. For $\delta < t_{s+3}$ the leftmost point of E is at $0' = t_s^*$, so this will play the role of 0 (see Figure 14). To match the original proof, let us also set $t'_k = 0' + t_{k+1}$ if $k \geq s-1$ and $t'_k = t_k$ if $k < s-1$. Note that $t'_{s-1} = t_{s-1}$, so the two sequences $\{t_k\}$ and $\{t'_k\}$ separate only from the index $k = s$. Now let us copy Part I of the proof of Theorem 5.1 (this is the only part we need here) with the change that everywhere we write instead of $0, t_k, \mathcal{C}$ etc. the corresponding symbols $0', t'_k, \mathcal{C}'$ etc, except that we keep the constants D_n defined and used in that proof. In estimating

$$\overline{\mu\Big|_{[t'_{k+1}, t'_k] \setminus \mathcal{C}'_n}}([0', 0' + \delta]) \tag{5.63}$$

there is no change if $k \leq s-3$, for then the portions of the sets $\mathcal{C}_n$ and $\mathcal{C}'_n$ that the proof uses are the same. In a similar manner, if $k \geq s$, then the appropriate portions of $\mathcal{C}'_n$ (like $[t'_{k+1}, t'_k] \cap \mathcal{C}'_n$) is just a translated version of the appropriate portion of $\mathcal{C}_n$ (like $[t_{k+2}, t_{k+1}] \cap \mathcal{C}_n$); e.g.

$$[t'_{k+1}, t'_k] \cap \mathcal{C}'_n = \Big([t_{k+2}, t_{k+1}] \cap \mathcal{C}_n\Big) + t_s^*,$$

so again there is no difference in the proof. Finally, if $k = s-2$ or $k = s-1$, then all we have to do is to choose a different r_a (c.f. (5.42)). In these two cases let $r'_a = t_{s+3}$. With this choice (5.42) changes into

$$\frac{(1-\varepsilon)^4}{16} \leq \frac{r'_a}{t'_k} \leq 1,$$

but this does not cause any trouble (see the remark made after Lemma 5.8)). With these changes the proof of (5.42)–(5.49) goes over to $\mathcal{C}'_n$ with small changes, and as an analogue of (5.49) we obtain

$$\mu_{\mathcal{C}'_n}([0', 0' + \delta]) \leq \sqrt{\delta}\left(C_1 + C_2 D_{n-1} \sum_j \varepsilon_j^2\right),$$

with some constants C'_1, C'_2. Since we know that $\{D_n\}$ (which now corresponds to the bound $(1-\varepsilon)^4/16 \leq r \leq 1$ instead of $(1-\varepsilon)^2/4 \leq r \leq 1$ in (5.13)) is a bounded sequence, this provides the estimate (5.62).

The point is that the proof in Part I of Theorem 5.1 is local in the sense that in this proof we only have to look for $\mathcal{C}$ through the "window" which is the interior of $C_{r_a}(a)$, and we cannot distinguish $\mathcal{C}'$ from $\mathcal{C}$ in these "windows", so all the estimates work just as before.

■

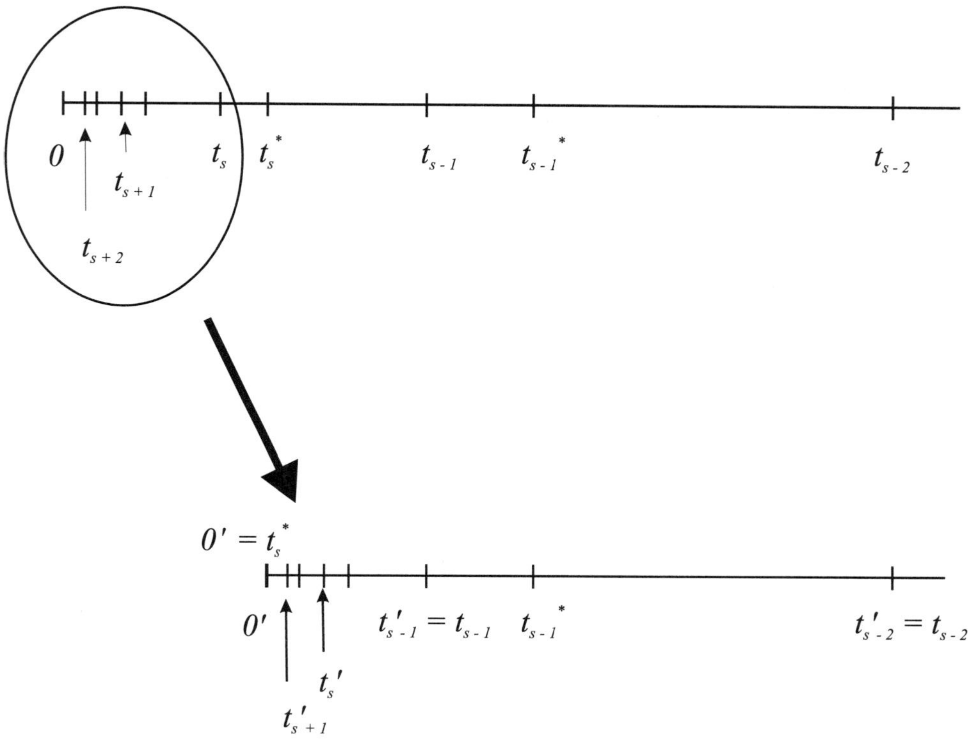

Figure 14:

6 Phargmén–Lindelöf type theorems

The following Phragmén–Lindelöf type theorem is classical (see e.g. [35], [29]): if f is analytic on $\mathbf{C} \setminus [0, \infty)$, has boundary values of modulus ≤ 1 on $[0, \infty)$ and

$$\lim_{|z|\to\infty} \frac{\log |f(z)|}{|z|^{1/2}} = 0, \tag{6.1}$$

then $|f(z)| \le 1$ for all z. The function $e^{\sqrt{z}}$ shows that this is sharp regarding the condition (6.1). In this section we consider the problem if there are analogous results when f is analytic on an infinitely connected domain, and how large boundary will force the above result to be valid.

Let $E \subset \mathbf{C}$ be a closed set such that $\mathbf{C} \setminus E$ is connected. We shall again measure the density of E by the function

$$\Theta_E(t) = |[0, t] \setminus \{r \,|\, C_r(0) \cap E \neq \emptyset\}|,$$

but this time we shall be interested in the behavior of Θ_E around $t = \infty$.

Theorem 6.1 *Let f be analytic on $\mathbf{C} \setminus E$ such that for each $z_0 \in E$*

$$\limsup_{z\to z_0,\ z\notin E} |f(z)| \le 1, \tag{6.2}$$

and

$$\lim_{|z|\to\infty,\ z\notin E} \frac{\log |f(z)|}{|z|^{1/2}} = 0. \tag{6.3}$$

If

$$\int_1^\infty \frac{\Theta_E(t)^2}{t^3} dt < \infty, \tag{6.4}$$

then these imply $|f(z)| \le 1$ for all z.

This theorem is sharp as is shown by

Theorem 6.2 *Let us suppose that $0 \le \Theta(t) \le t$ is an increasing function with*

$$\int_1^\infty \frac{\Theta(t)^2}{t^3} dt = \infty. \tag{6.5}$$

Then there are a closed set $E \subset [0, \infty)$ and an analytic function f on $\mathbf{C} \setminus E$ with the properties (6.2) *and* (6.3)*, and yet $|f(z)| > 1$ for some z.*

Theorem 6.1 can be considered to be the case $\alpha = 0$ of the next result, in which we use again the set (c.f. (4.9))

$$E_\alpha = \{r \,|\, |C_r(0) \cap E| \ge 2\pi r\alpha\}.$$

Theorem 6.3 *Let $0 < \alpha < 1$, and let f be an analytic function on $\mathbf{C} \setminus E$ with the properties* (6.2) *and*

$$\lim_{|z|\to\infty,\ z\notin E} \frac{\log |f(z)|}{|z|^{1/2(1-\alpha)}} = 0. \tag{6.6}$$

If

$$\int_1^\infty \frac{\Theta_{E_\alpha}(t)^2}{t^3} dt < \infty, \tag{6.7}$$

then these imply $|f(z)| \le 1$ for all z.

Since $\log |f(z)|$ is subharmonic, both of Theorems 6.1 and 6.3 immediately follow from the following version, in which we set $\Theta_{E_0}(t) = \Theta_E(t)$.

Theorem 6.4 *Let $0 \leq \alpha < 1$ and let u be a subharmonic function on $\mathbf{C} \setminus E$ such that u is bounded from above on every set $C_R(0) \setminus E$, $R > 0$, for each $z_0 \in E$*

$$\limsup_{z \to z_0,\ z \notin E} |u(z)| \leq 0, \tag{6.8}$$

and

$$\limsup_{|z| \to \infty,\ z \notin E} \frac{u(z)}{|z|^{1/2(1-\alpha)}} \leq 0. \tag{6.9}$$

If (6.7) is satisfied, then these imply $u(z) \leq 0$ for all z.

We shall be content to prove only for the case $\alpha = 0$ that regarding the finiteness of the integral in (6.4) this result is sharp. Naturally the sharpness of this result follows from Theorem 6.2, however the proof of Theorem 6.2 is very involved, and in any case it goes through the reasoning what we shall present for

Theorem 6.5 *Let us suppose that $0 \leq \Theta(t) \leq t$ is an increasing function with the property (6.5). Then there are a closed set $E \subset [0, \infty)$ and a harmonic function u on $\mathbf{C} \setminus E$ with the properties that u is bounded on every set $C_R(0) \setminus E$, $R > 0$, for each $z_0 \in E$ the property (6.8) holds and so does*

$$\lim_{|z| \to \infty,\ z \notin E} \frac{u(z)}{|z|^{1/2}} = 0, \tag{6.10}$$

and yet $u(z) > 0$ for some z.

In this section we prove Theorems 6.4 and 6.5. The proof of Theorem 6.2 will be presented in the next section.

Proof of Theorem 6.4. This is a standard argument, but for completeness let us present it here.

Fix $z_0 \in \mathbf{C} \setminus E$. For large R and $|z| < R$ we get from Theorem 4.3

$$\begin{aligned} \omega(z, C_R(0), \Delta_R(0) \setminus E) &= \omega(z/R, C_1(0), \Delta_1(0) \setminus (E/R)) \\ &\leq C(|z|/R)^{1/2(1-\alpha)} \exp\left(C \int_{|z|/R}^{1} \frac{\Theta^2_{E_\alpha/R}(u)}{u^3} du \right), \end{aligned}$$

where C is a constant depending only on α, and where E/R denotes the image of E under the contraction $z \to z/R$. Clearly $\Theta_{E_\alpha/R}(v/R) = \Theta_{E_\alpha}(v)/R$, hence the substitution $u = v/R$ shows in conjunction with (6.7) that there is a constant C such that

$$\omega(z, C_R(0), \Delta_R(0) \setminus E) \leq C \left(\frac{|z|}{R} \right)^{1/2(1-\alpha)}. \tag{6.11}$$

Thus, if $\varepsilon > 0$ then we get from the conditions (6.8) – (6.9) for sufficiently large R

$$\limsup_{z \to z',\ z \notin E} u(z) \leq \varepsilon R^{1/2(1-\alpha)} \omega(z', C_R(0), \Delta_R(0) \setminus E)$$

for all $z' \in \partial(\Delta_R(0)\setminus E)$, and so the subharmonicity of u and the maximum principle gives

$$u(z) \leq \varepsilon R^{1/2(1-\alpha)}\omega(z, C_R(0), \Delta_R(0) \setminus E)$$

for all $z \in C_R(0) \setminus E$. This combined with (6.11) yields

$$u(z) \leq C\varepsilon|z|^{1/2(1-\alpha)}.$$

Setting here $z = z_0$ and letting $\varepsilon \to 0$ we obtain $u(z_0) \leq 0$ as was claimed.

■

Proof of Theorem 6.5. Since (6.5) says something about Θ around infinity, we may assume $\Theta(t) = t$ for $t \in [0, 2]$ (otherwise we would have to consider a translate of a portion of the set to be constructed below). Let

$$E = \bigcup_{k=0}^{\infty}\left[2^k + \frac{1}{4}\left(\Theta\left(2^k\right) - \Theta\left(2^{k-1}\right)\right), 2^{k+1}\right]. \tag{6.12}$$

This is the analogue of the set (3.10), and exactly as it was proven for (3.10) we get that $\Theta_E(t) \leq \Theta(t)$ for all $t \geq 0$. It also follows from Theorem 3.1, (3.5) (apply the mapping $t \to t/R$ as in the preceding proof) that

$$R^{1/2}\omega(0, C_R(0), \Delta_R(0) \setminus E) \to \infty, \qquad \text{for } R \to \infty. \tag{6.13}$$

Consider the functions

$$\omega_n(z) = \frac{\omega(z, C_n(0), \Delta_n(0) \setminus E)}{\omega(0, C_n(0), \Delta_n(0) \setminus E)}.$$

These are nonnegative harmonic functions (on their domains) and take the value 1 at the origin, so by Harnack's principle they are asymptotically bounded on compact subsets of $\mathbf{C} \setminus E$ in the sense that if $K \subset \mathbf{C} \setminus E$ is compact, then there is an N and a bound M such that $\omega_n(z)$ is defined on K for $n > N$ and is of modulus at most M. Hence we can select a locally uniformly converging subsequence from ω_n, say $\omega_{n_i}(z) \to \omega(z)$ uniformly on each compact subset of $\mathbf{C} \setminus E$. But this latter set contains full circles of arbitrary large radii, so we can infer from $\omega_n(z) = 0$ for $z \in \Delta_R(0) \cap E$ and from the maximum principle that the convergence is actually uniform on compact subsets of $\mathbf{C}$, and in particular $\omega(z) = 0$ for $z \in E$.

We are going to prove that $\omega(z)/|z|^{1/2} \to 0$ as $z \to \infty$, which will complete the proof since then E and $u = \omega$ are appropriate in the theorem (note that ω is harmonic outside E and $\omega(0) = 1$). It is a consequence of Beurling's theorem [27, Sec. IV.5.4] that each ω_n takes its largest value at $-r$ on every circle $C_r(0)$ $(r < n)$, which property is then inherited by ω, i.e. $\omega(z) \leq \omega(-|z|)$. Since $\mathbf{C}\setminus[1, \infty) \subseteq \mathbf{C}\setminus E$, it follows from Harnack's inequality that

$$\omega(-re^{it}) \geq \frac{1}{6}\omega(-r) \tag{6.14}$$

for $t \in [-\pi/2, \pi/2]$. Now we use that if

$$C_r^-(0) = \{-re^{it} \,|\, t \in [-\pi/2, \pi/2]\}$$

is the left half of the circle $C_r(0)$, then

$$\omega(0, C_r^-(0), \Delta_r(0) \setminus E) \geq \frac{1}{2}\omega(0, C_r(0), \Delta_r(0) \setminus E). \tag{6.15}$$

Taking this for granted the proof can be completed as follows. If $\widehat{\delta_0}$ denotes the balayage measure of δ_0 out of $\Delta_r(0) \setminus E$, then on the one hand

$$\omega(0) = \int_{\partial(\Delta_r(0)\setminus E)} \omega \, d\widehat{\delta_0} = \int_{C_r(0)} \omega \, d\widehat{\delta_0}$$

because ω vanishes on E, and on the other hand

$$\int_{C_r^-(0)} d\widehat{\delta_0} = \widehat{\delta_0}(C_r^-(0)) = \omega(0, C_r^-(0), \Delta_r(0) \setminus E) \geq \frac{1}{2}\omega(0, C_r(0), \Delta_r(0) \setminus E).$$

Thus, we obtain from (6.14)

$$1 = \omega(0) = \int_{C_r(0)} \omega \, d\widehat{\delta_0} \geq \frac{1}{6}\omega(-r) \int_{C_r^-(0)} d\widehat{\delta_0} \geq \frac{1}{12}\omega(-r)\omega(0, C_r(0), \Delta_r(0) \setminus E),$$

and then (6.13) shows that, indeed, for $r = |z|$

$$\frac{\omega(z)}{|z|^{1/2}} \leq \frac{\omega(-r)}{r^{1/2}} \leq \frac{2}{c_0} \frac{1}{r^{1/2}\omega(0, C_r(0), \Delta_r(0) \setminus E)} \to 0$$

as $r \to \infty$.

This completes the proof pending the verification of (6.15). But that is easy: as we have just used it,

$$\omega(0, C_r^-(0), \Delta_r(0) \setminus E) = \widehat{\delta_0}(C_r^-(0)), \quad \text{and} \quad \omega(0, C_r(0), \Delta_r(0) \setminus E) = \widehat{\delta_0}(C_r(0)).$$

If $\overline{\delta_a}$ is the balayage measure of δ_a out of $\Delta_r(0)$, then $\overline{\delta_0}$ can be obtained in two steps: first take the balayage out of $\Delta_r(0) \setminus E$ to obtain $\widehat{\delta_0}$, and then take the balayage of this measure out of $\Delta_r(0)$, i.e. for any measurable subset A of the circle $C_r(0)$

$$\overline{\delta_0}(A) = \widehat{\delta_0}(A) + \int_{E\cap[0,r]} \overline{\delta_a}(A) \, d\widehat{\delta_0}(a). \tag{6.16}$$

In particular,

$$\widehat{\delta_0}(C_r^-(0)) = \frac{1}{2} - \int_{E\cap[0,r]} \overline{\delta_a}(C_r^-(0)) \, d\widehat{\delta_0}(a). \tag{6.17}$$

But $\overline{\delta_a}$ is given by the Poisson kernel

$$d\overline{\delta_a}(re^{it}) = \frac{1}{2\pi} \frac{r^2 - a^2}{r^2 - 2ra\cos t + a^2} dt,$$

and since for $a > 0$ this latter expression increases for $t \in [-\pi, 0]$ and decreases for $t \in [0, \pi]$, we get

$$\int_{E\cap[0,r]} \overline{\delta_a}(C_r^-(0)) \, d\widehat{\delta_0}(a) \leq \frac{1}{2} \int_{E\cap[0,r]} \overline{\delta_a}(C_r(0)) \, d\widehat{\delta_0}(a).$$

Substituting this into (6.17) we can infer

$$\widehat{\delta_0}(C_r^-(0)) \geq \frac{1}{2} - \frac{1}{2}\int_{E\cap[0,r]} \overline{\widehat{\delta_a}}(C_r(0))\, d\widehat{\delta_0}(a),$$

and another application of (6.16) shows that here the right hand side equals

$$\frac{1}{2}\widehat{\delta_0}(C_r(0)) = \frac{1}{2}\omega(0, C_r(0), \Delta_r(0)\setminus E),$$

and (6.15) follows.

∎

6.1 Proof of Theorem 6.2

The general direction of the proof is clear: let E be the set from the preceding proof, E_n the union of the first n subintervals of E and let

$$g_n(z) = \frac{g_{\mathbf{C}\setminus E_n}(z)}{g_{\mathbf{C}\setminus E_n}(0)}$$

where $g_{\mathbf{C}\setminus E_n}$ is the Green function for the complement of E_n. Set

$$h_n(z) = \exp(g_n(z) + iv_n(z)),$$

where v_n is the harmonic conjugate of g_n. Then it is clear that $|h_n(z)| = 1$ on E_n but $|h_n(0)| = e > 1$, and as in the previous proof, one can verify that for the limit f of the functions h_n (provided it exists) condition (6.3) is satisfied. However, the h_n's are multivalent. To get single valued h_n's this way the following condition has to be met: since v_n gives the argument of f_n; as we circle a subinterval of E once, the total change in v_n must be a multiple of 2π, i.e. if I is a subinterval of E, then we must have

$$\oint_I (v_n)_x(x)dx = k_{n,I}2\pi \tag{6.18}$$

with some integer $k_{n,I}$. By the Cauchy-Riemann equations this takes the form

$$\oint_I (g_n)_y(x)dx = -k_{n,I}2\pi. \tag{6.19}$$

But g_n is a constant $(= 1/g_{\mathbf{C}\setminus E_n}(0))$ multiple of the Green function for the set E_n, so the normal derivative $(g_n)_y$ gives a constant times the equilibrium measure of the set, more precisely

$$\frac{d\mu_{E_n}(t)}{dt} = \frac{1}{\pi}\frac{\partial g_{\mathbf{C}\setminus E_n}(t)}{\partial \mathbf{n}},$$

which yields

$$\oint_I (g_n)_y(x)dx = 2\pi\frac{\mu_{E_n}(I)}{g_{\mathbf{C}\setminus E_n}(0)}.$$

Thus, we must have

$$\frac{\mu_{E_n}(I)}{g_{\mathbf{C}\setminus E_n}(0)} = -k_{n,I} \tag{6.20}$$

for all subintervals I of E_n.

This is quite difficult to achieve, but we shall follow this general direction with several modifications, one of which will be to achieve for all subintervals I of E_n

$$D_n \mu_{E_n}(I) = k^*_{n,I} \tag{6.21}$$

with some integers $k^*_{n,I}$ and some constants D_n for which

$$d < D_n / g_{\mathbf{C} \setminus E_n}(0) < D, \qquad n = 1, 2, \ldots \tag{6.22}$$

(and then just work with $g_n(z) = D_n g_{\mathbf{C}\setminus E_n}(z)$ as above). To ensure the first condition (6.21) we have to move the endpoints of the intervals in E_n by a sufficiently large amount so that we hit integer values on the left of (6.21), but we should do it in a fashion that we should not ruin (6.22). This balancing needs some very careful estimates of equilibrium measures and balayage measures. When we move n endpoints of the n subintervals $I_1, \ldots, I_n$ of E_n by some amounts $x_1, \ldots, x_n$, then μ_{E_n} will depend on all the variables $x_1, \ldots, x_n$, so $D_n \mu_{E_n}(I_j)$ will be functions of n variables, and our aim is to choose the variables so that all these functions of n variables take integer values. Thus, we face the problem to get integer values for the vector valued function

$$(x_1, \ldots, x_n) \to (D_n \mu_{E_n}(I_1), \ldots, D_n \mu_{E_n}(I_n))$$

(from some subset of $\mathbf{R}^n$ into $\mathbf{R}^n$). What we shall show is that the image of an appropriate cube contains the boundary of a sufficiently large "curved cube", and then the fact that there are integer points in the range will follow from Brouwer's fixed point theorem.

The actual construction will be much more complicated, but the main direction will be as we have just outlined.

Since it will be very important in the proof that certain constants do not depend on anything, in the proof below we calculate all the absolute constants. This is somewhat awkward for the constants may be quite large, but as we have said, it helps to verify the main ideas of the proof.

Choice of the numbers θ_j

We set

$$\theta^*_j = \frac{\Theta(2^{j-2}) - \Theta(2^{j-3})}{8 \cdot 2^j} \tag{6.23}$$

for $j = 1, 2, \ldots$. We can easily get from Lemma 11.3 (apply a change of variables) that the assumption on Θ implies $\sum_j (\theta^*_j)^2 = \infty$. Therefore, no matter how small the number $0 < \theta < 1/16$ is, we may choose $0 \le \theta_j \le \theta^*_j$ with the properties

$$\theta_j \le \theta < 1/16, \qquad \sum_j \theta_j^2 = \infty, \qquad \sum_j \theta_j^3 < \theta, \tag{6.24}$$

and we may also assume that for all j

$$\theta_j \ge 1/j \quad \text{or} \quad \theta_j = 0. \tag{6.25}$$

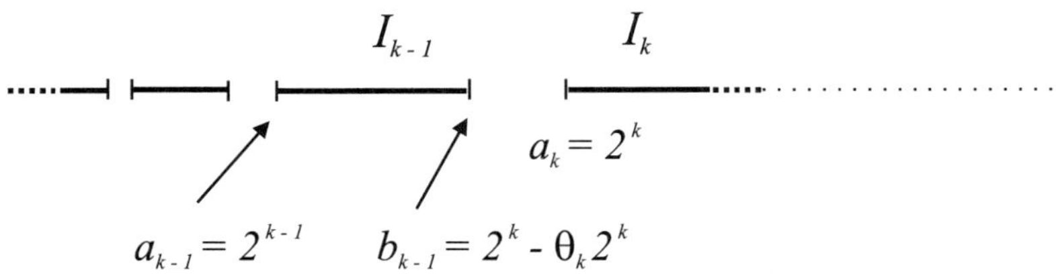

Figure 15:

In particular, we definitely have $\theta_1 = \ldots, \theta_{16} = 0$.

The number θ appearing in (6.24) will be chosen later. However, in the first part of the proof we shall only use that $\theta_j < 1/16$, and so the constants appearing in this part will not depend on the choice of θ.

Let $a_k = 2^k$, $b_k = 2^{k+1} - \theta_{k+1}2^{k+1}$, $I_k = [a_k, b_k]$, $k = 0, 1, \ldots$ and

$$K_n = \bigcup_{k=0}^{n-1} I_k = \bigcup_{k=0}^{n-1} [2^k, 2^{k+1} - \theta_{k+1}2^{k+1}] \tag{6.26}$$

(see Figure 15). By (2.17) the density ω_{K_n} of the equilibrium measure μ_{K_n} of K_n has the form for $t \in K_n$,

$$\omega_{K_n}(t) = \frac{1}{\pi} \frac{1}{\sqrt{|t-1||t-b_{n-1}|}} \prod_{j=1}^{n-1} \frac{|t - \xi_{n,j}|}{\sqrt{|t-b_{j-1}||t-a_j|}}, \tag{6.27}$$

where $\xi_{n,j}$, $j = 1, \ldots, n-1$ are the solutions of the systems of equations

$$\int_{b_{j-1}}^{a_j} \omega_{K_n}(t)\mathrm{sign}(t - \xi_{n,j})dt = 0 \qquad j = 1, \ldots, n-1.$$

Here we used the understanding that if $\theta_j = 0$, then there is no gap between the intervals I_{j-1} and I_j and in this case $b_{j-1} = \xi_{n,j} = a_j$, but otherwise we have $b_{j-1} < \xi_{n,j} < a_j$.

First we estimate $\omega_{K_n}(t)$ for $t \in [2^k, 2^{k+1}]$.

Estimation of the density ω_{K_n} of the equilibrium measure

The expression (6.27) for $\omega_{K_n}(t)$ contains a product $\prod_{j=1}^{n-1}$. In what follows for $t \in [2^k, 2^{k+1}]$, $0 \le k \le n-1$ we shall separately estimate the corresponding partial products $\prod_{j=1}^{k-1}$ and $\prod_{j=k+2}^{n-1}$.

Set $M_{n,n-2} = M_{n,n-1} = 1$ and

$$M_{n,k} = \prod_{j=k+2}^{n-1} \frac{\xi_{n,j}}{\sqrt{b_{j-1}a_j}}, \tag{6.28}$$

for $k < n-2$. Then

$$\frac{M_{n,k}}{M_{n,k+1}} = \frac{\xi_{k+2}}{\sqrt{b_{k+1}a_{k+2}}} \le \sqrt{\frac{a_{k+2}}{b_{k+1}}} \le \sqrt{1 + \frac{\theta_{k+2}2^{k+2}}{2^{k+1}}} \le 1 + \theta_{k+2} \le 1 + \frac{1}{16},$$

and similar reasoning shows that the fraction on the left is at least as large as $1/(1+\theta_{k+2}) \ge 1/(1+1/6)$. Thus

$$\frac{16}{17} \le \frac{M_{n,k}}{M_{n,k+1}} \le \frac{17}{16} \tag{6.29}$$

for all k, and together with these we also have

$$\left(\frac{16}{17}\right)^{|k-j|} \le \frac{M_{n,k}}{M_{n,j}} \le \left(\frac{17}{16}\right)^{|k-j|}. \tag{6.30}$$

If $t \in [2^k, 2^{k+1}]$, then

$$\prod_{j=1}^{k-1} \frac{|t-\xi_{n,j}|}{\sqrt{|t-b_{j-1}||t-a_j|}} \le \prod_{j=1}^{k-1} \sqrt{\frac{|t-b_{j-1}|}{|t-a_j|}} = \prod_{j=1}^{k-1} \sqrt{1 + \frac{a_j - b_{j-1}}{t-a_j}}$$

$$\le \prod_{j=1}^{k-1} \sqrt{1 + \frac{\theta_j 2^j}{2^{k-1}}} \le \prod_{j=1}^{k-1} \left(1 + 2^{j-k+1}\right)^{1/2} \le \exp\left(\sum_{j=1}^{k-1} 2^{j-k}\right) \le e,$$

and similar reasoning shows that

$$\prod_{j=1}^{k-1} \frac{|t-\xi_{n,j}|}{\sqrt{|t-b_{j-1}||t-a_j|}} \ge \frac{1}{e}.$$

In the following estimates we use that for $\alpha > \beta$ the function $|t-\alpha|/|t-\beta|$ is increasing on the intervals $(-\infty,\beta)$ and (α,∞), and therefore its reciprocal $|t-\beta|/|t-\alpha|$ is decreasing on these intervals. Write

$$\prod_{j=k+2}^{n-1} \frac{|t-\xi_{n,j}|}{\sqrt{|t-b_{j-1}||t-a_j|}} = \frac{\sqrt{|t-b_{n-1}|}}{\sqrt{|t-b_{k+1}|}} \prod_{j=k+2}^{n-1} \frac{|t-\xi_{n,j}|}{\sqrt{|t-a_j||t-b_j|}}.$$

In the last product every factor is decreasing on $(-\infty, \xi_{k+2}) = (-\infty, 2^{k+2})$, hence this product is at most as large as its value at $t = 0$, which is

$$M_{n,k}\frac{\sqrt{b_{k+1}}}{\sqrt{b_{n-1}}}.$$

Thus,

$$\prod_{j=k+2}^{n-1} \frac{|t-\xi_{n,j}|}{\sqrt{|t-b_{j-1}||t-a_j|}} \le \frac{\sqrt{2^n}}{\sqrt{2^k}} M_{n,k} \frac{\sqrt{b_{k+1}}}{\sqrt{b_{n-1}}} \le \frac{\sqrt{2^n}}{\sqrt{2^k}} M_{n,k} \frac{\sqrt{2^{k+2}}}{\sqrt{2^{n-1}}} \le 4M_{n,k}.$$

On the other hand,

$$\prod_{j=k+2}^{n-1} \frac{|t-\xi_{n,j}|}{\sqrt{|t-b_{j-1}||t-a_j|}} = \frac{\sqrt{|t-\xi_{n,k+2}|}}{\sqrt{|t-a_{n-1}|}} \times$$

$$\times \frac{\sqrt{|t-\xi_{n,k+2}|}}{\sqrt{|t-b_{k+1}|}} \prod_{j=k+3}^{n-1} \frac{|t-\xi_{n,j}|}{\sqrt{|t-a_{j-1}||t-b_{j-1}|}},$$

and here after the $\times$ sign every factor increases on the interval $(-\infty, b_{k+1})$, therefore we estimate this product from below if we take its value at the origin, which is

$$M_{n,k}\frac{\sqrt{a_{n-1}}}{\sqrt{\xi_{n,k+2}}}.$$

We get this way

$$\prod_{j=k+2}^{n-1} \frac{|t-\xi_{n,j}|}{\sqrt{|t-b_{j-1}||t-a_j|}} \geq \frac{\sqrt{2^k}}{\sqrt{2^n}} M_{n,k}\frac{\sqrt{a_{n-1}}}{\sqrt{\xi_{n,k+2}}} \geq \frac{\sqrt{2^k}}{\sqrt{2^n}} M_{n,k}\frac{\sqrt{2^{n-1}}}{\sqrt{2^{k+2}}} \geq \frac{M_{n,k}}{4}.$$

All these give for $t \in [2^k, 2^{k+1}]$

$$\omega_{K_n}(t) \leq 4eM_{n,k}\frac{1}{\sqrt{|t-1||t-b_{n-1}|}} \frac{|t-\xi_{n,k}|}{\sqrt{|t-b_{k-1}||t-a_k|}} \frac{|t-\xi_{n,k+1}|}{\sqrt{|t-b_k||t-a_{k+1}|}}, \tag{6.31}$$

and from below we have a similar estimate

$$\omega_{K_n}(t) \geq \frac{1}{4e} M_{n,k} \cdots \tag{6.32}$$

where $\cdots$ denotes the product on the right of the previous formula.

These inequalities hold for $0 < k < n-1$. If $k = 0$ and $k = n-1$ then similar formulae are true, just for $k = n-1$ the last ratio is missing on the right hand sides, while for $k = 0$ the second ratio is missing (but actually for $k = 0$ the last factor is also identically 1 because $b_1 = a_2 = \xi_1$).

For the first ratio in (6.31) we have for $t \in [2^k, 2^{k+1}]$, $0 < k < n-1$

$$\frac{1}{\sqrt{2^{k+1}}\sqrt{2^n}} \leq \frac{1}{\sqrt{|t-1||t-b_{n-1}|}} \leq \frac{1}{\sqrt{2^{k-1}}\sqrt{2^{n-1}}},$$

i.e.

$$\frac{1}{2}\frac{1}{\sqrt{2^k}\sqrt{2^n}} \leq \frac{1}{\sqrt{|t-1||t-b_{n-1}|}} \leq \frac{2}{\sqrt{2^k}\sqrt{2^n}}, \tag{6.33}$$

but to estimate the other two ratios we need information on where the numbers $\xi_{n,k}$ are located in the intervals $[b_{k-1}, a_k]$.

Location of $\xi_{n,k}$

Let $t \in (b_k, a_{k+1})$. Then

$$\frac{1}{2} \leq \frac{|t-\xi_{n,k}|}{\sqrt{|t-b_{k-1}||t-a_k|}} \leq 2. \tag{6.34}$$

Since the equality

$$\int_{b_k}^{a_{k+1}} \omega_{K_n}(t)\operatorname{sign}(t-\xi_{n,k+1})dt = 0$$

holds, we obtain from (6.31) – (6.34) that

$$\left(-\frac{16eM_{n,k}}{\sqrt{2^k}\sqrt{2^n}}\int_{b_k}^{\xi_{n,k+1}} + \frac{M_{n,k}}{16e\sqrt{2^k}\sqrt{2^n}}\int_{\xi_{n,k+1}}^{a_{k+1}}\right)\frac{|t-\xi_{n,k+1}|}{\sqrt{|t-b_k||t-a_{k+1}|}}dt < 0,$$

and at the same time

$$\left(-\frac{M_{n,k}}{16e\sqrt{2^k}\sqrt{2^n}}\int_{b_k}^{\xi_{n,k+1}} + \frac{16eM_{n,k}}{\sqrt{2^k}\sqrt{2^n}}\int_{\xi_{n,k+1}}^{a_{k+1}}\right)\frac{|t-\xi_{n,k+1}|}{\sqrt{|t-b_k||t-a_{k+1}|}}dt > 0. \quad (6.35)$$

If we define ξ by the equation

$$\xi_{n,k+1} = b_k + \xi(a_{k+1} - b_k),$$

then the first of the preceding two inequalities takes the form

$$\left(-16e\int_0^{\xi} + \frac{1}{16e}\int_{\xi}^{1}\right)\frac{|t-\xi|}{\sqrt{t(1-t)}}dt < 0.$$

Thus,

$$((16e)^2+1)\int_0^{\xi}\frac{|t-\xi|}{\sqrt{t(1-t)}}dt \ge \int_0^1 \frac{|t-\xi|}{\sqrt{t(1-t)}}dt \ge \frac{1}{4}\int_0^{1/4}\frac{1}{\sqrt{t}}dt = \frac{1}{4}.$$

Here either $\xi \ge 1/2$, or the integrand on the left is increasing on $[0,\xi]$, in which case we obtain

$$((16e)^2+1)\int_0^{\xi/2}\frac{\xi/2}{\sqrt{t}\sqrt{1/2}}dt \ge \frac{1}{8},$$

i.e.

$$((16e)^2+1)\xi^{3/2} \ge \frac{1}{8}.$$

Thus, we have verified that in any case

$$\xi \ge \left(\frac{1}{8((16e)^2+1)}\right)^{2/3} \ge \frac{1}{700},$$

that is

$$\xi_{n,k+1} \ge b_k + \frac{\theta_{k+1}2^{k+1}}{700}. \quad (6.36)$$

Using (6.35), the same calculation gives that

$$\xi_{n,k+1} \le a_{k+1} - \frac{\theta_{k+1}2^{k+1}}{700}. \quad (6.37)$$

Returning to the estimate of ω_{K_n}

After this we return to the estimate of the density of the equilibrium measure. First consider the case when $t \in [2^k, 2^{k+1}]$ also satisfies

$$a_k + \theta_k 2^k \le t \le b_k - \theta_{k+1} 2^{k+1}.$$

For such t the last two ratios on the right of (6.31) – (6.32) both lie between $1/\sqrt{2}$ and $\sqrt{2}$, therefore for such values of t we get from (6.31) – (6.32) and (6.33) that

$$\frac{M_{n,k}}{16e\sqrt{2^k}\sqrt{2^n}} \le \omega_{K_n}(t) \le \frac{16eM_{n,k}}{\sqrt{2^k}\sqrt{2^n}}. \tag{6.38}$$

Next, let

$$a_k \le t \le a_k + \theta_k 2^k.$$

Then similarly as before we derive from (6.31) and (6.33)

$$\omega_{K_n}(t) \le \frac{16eM_{n,k}}{\sqrt{2^k}\sqrt{2^n}} \frac{\sqrt{\theta_k 2^k}}{\sqrt{|t-a_k|}} = \frac{16eM_{n,k}}{\sqrt{2^n}} \frac{\sqrt{\theta_k}}{\sqrt{|t-a_k|}}. \tag{6.39}$$

To get the corresponding lower estimate we use (6.32) and (6.33) combined with (6.36) – (6.37) (with $k+1$ replaced by k):

$$\omega_{K_n}(t) \ge \frac{M_{n,k}}{16e} \frac{1}{\sqrt{2^k}\sqrt{2^n}} \frac{\sqrt{\theta_k 2^k}}{700\sqrt{|t-a_k|}} \ge \frac{1}{4\cdot 10^4} \frac{M_{n,k}}{\sqrt{2^n}} \frac{\sqrt{\theta_k}}{\sqrt{|t-a_k|}}. \tag{6.40}$$

Similar inequalities are true for

$$b_k - \theta_{k+1} 2^{k+1} \le t \le b_k,$$

just replace $\sqrt{\theta_k/|t-a_k|}$ by $\sqrt{\theta_{k+1}/|t-b_k|}$.

Since $b_0 = a_1 = \xi_1$, completely analogous argument gives that for $t \in [1,2]$ we have

$$\frac{1}{16e} \frac{M_{n,0}}{\sqrt{t-1}\sqrt{2^n}} \le \omega_{K_n}(t) \le 16e \frac{M_{n,0}}{\sqrt{t-1}\sqrt{2^n}}. \tag{6.41}$$

Mass of the equilibrium measure on different parts of I_k

Now we are already estimate the total mass of the equilibrium measure μ_{K_n} on the interval I_k.

First of all, (6.38) gives that

$$\frac{M_{n,k}\sqrt{2^k}}{32e\sqrt{2^n}} \le \mu_{K_n}([a_k + \theta_k 2^k, b_k - \theta_{k+1} 2^{k+1}]) \le 16e \frac{M_{n,k}\sqrt{2^k}}{\sqrt{2^n}}, \tag{6.42}$$

while integrating the inequality in (6.39) from a_k to $a_k + \theta_k 2^k$ we get

$$\mu_{K_n}([a_k, a_k + \theta_k 2^k]) \le 32e \frac{M_{n,k}\theta_k\sqrt{2^k}}{\sqrt{2^n}}. \tag{6.43}$$

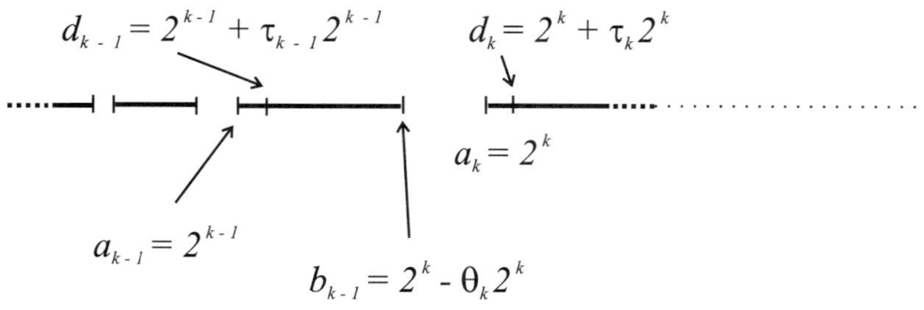

Figure 16:

In a similar manner by integrating the inequality in (6.40) from a_k to $a_k + \theta_k 2^k/2$ we get

$$\mu_{K_n}([a_k, a_k + \theta_k 2^k/2]) \geq \frac{M_{n,k}\theta_k\sqrt{2^k}}{4 \cdot 10^5\sqrt{2^n}}. \tag{6.44}$$

The same estimate holds for the mass on the interval $[b_k - \theta_{k+1}2^{k+1}, b_k]$, just replace θ_k by θ_{k+1}. Taking into account that $\theta_k \leq 1/16$, these give

$$\frac{M_{n,k}\sqrt{2^k}}{100\sqrt{2^n}} \leq \mu_{K_n}(I_k) \leq 100\frac{M_{n,k}\sqrt{2^k}}{\sqrt{2^n}}. \tag{6.45}$$

Next we need to estimate balayage measures.

Balayage out of a subinterval of I_k

Let $d_k = a_k + \tau_k 2^k$, where $0 \leq \tau_k \leq \theta_k$ are some numbers to be chosen below (see Figure 16). We shall get our set E_n by which we will prove the theorem by moving the left endpoints a_k of the intervals I_k to the right by some amount that does not exceed $\tau_k 2^k$, i.e. the left endpoints of subintervals of E_n lie in the intervals $[a_k, b_k]$. Note also that if $\theta_k = 0$, then $b_{k-1} = a_k = d_k$, so in this case this point actually will not be an endpoint for a subinterval of E_n.

Thus, the set E_n will be part of K_n, and its equilibrium measure is obtained from that of K_n by taking the balayage of the latter one onto E_n. Therefore, first we give estimates on balayage measures out of the intervals $[a_k, d_k]$.

Let $\mathrm{Bal}\Big(\nu, A\Big)$ denote the balayage of a measure ν onto the compact set A.

Let $1 \leq k \leq n-1$ be an index for which $\theta_k > 0$. For $s \in (b_{k-1}, d_k) = (2^k - \theta_k 2^k, 2^k + \tau_k 2^k)$ we estimate the density of the balayage of the Dirac mass δ_s placed at s. Let $b_k \leq b \leq b_{n-1}$ be any number, later we shall need both the choices $b = b_k$ and $b = b_{n-1}$. By (2.25) we have for $t \in [1, b_{k-1}] \cup [d_k, b]$

$$\frac{d\mathrm{Bal}\Big(\delta_s, [1, b_{k-1}] \cup [d_k, b]\Big)(t)}{dt} = H(t)$$

with

$$H(t) = H_{k,b,s}(t) = \frac{\sqrt{|s-1||s-b_{k-1}||s-d_k||s-b|}}{\pi\sqrt{|t-1||t-b_{k-1}||t-d_k||t-b|}}\frac{|t-\zeta|}{|s-\zeta|}\frac{1}{|t-s|},$$

where $\zeta = \zeta_{k,b,s}$ is determined by the equation

$$\left(\int_{-\infty}^{1} + \int_{b}^{\infty}\right) H(t)\mathrm{sign}(t-\zeta)dt = 0.$$

Here either $\zeta \leq 1$ or $\zeta \geq b \geq b_k = 2^{k+1} - \theta_{k+1}2^{k+1}$, and in either case for $t \in [1, 3 \cdot 2^{k-2}]$ we get

$$\frac{|t-\zeta|}{|s-\zeta|} \leq 3$$

and

$$\frac{|s-b|}{|t-b|} \leq 1.$$

Therefore for such values of t we have

$$H(t) \leq \frac{\sqrt{2^{k+1}(2\theta_k 2^k)|s-d_k|}}{\pi\sqrt{|t-1|2^{k-3}2^{k-2}}} \frac{1}{2^{k-2}} 3 \leq 50 \frac{\sqrt{\theta}_k\sqrt{|s-d_k|}}{\sqrt{|t-1|}2^k}. \tag{6.46}$$

In particular, if $t \in [1, 3 \cdot 2^{k-2}]$ then for $s \in [b_{k-1}, d_k]$

$$H(t) \leq 50\sqrt{2} \frac{\theta_k}{\sqrt{|t-1|}\sqrt{2^k}}, \tag{6.47}$$

while for $s \in [a_k, d_k]$

$$H(t) \leq 50 \frac{\sqrt{\theta}_k\sqrt{\tau_k 2^k}}{\sqrt{|t-1|}2^k}. \tag{6.48}$$

On the other hand, if $t \in [2^{k-1}, b_{k-1}]$ and $s \in [a_k, d_k]$ then

$$\frac{|t-\zeta|}{|s-\zeta|} \geq \frac{1}{3}$$

and

$$\frac{|s-b|}{|t-b|} \geq \frac{1}{2},$$

so for such t and s we have

$$H(t) \geq \frac{\sqrt{2^{k-1}(\theta_k 2^k)|s-d_k|}}{3\sqrt{2}\pi\sqrt{2^k|t-b_{k-1}||t-d_k|}} \frac{1}{|t-s|},$$

which for $t \in [2^{k-1}, b_{k-1} - \theta_k 2^k]$ gives

$$H(t) \geq \frac{\sqrt{\theta_k 2^k |s-d_k|}}{18\sqrt{2}\pi|t-b_{k-1}|^2}. \tag{6.49}$$

By integration this from $3 \cdot 2^{k-2}$ to $b_{k-1} - \theta_k 2^k$ we obtain for the balayage measure of the right half of the interval I_k:

$$\mathrm{Bal}\Big(\delta_s, [1, b_{k-1}] \cup [d_k, b]\Big)([3 \cdot 2^{k-2}, b_{k-1}]) \geq \frac{\sqrt{|s-d_k|}}{30\pi\sqrt{\theta_k 2^k}}.$$

Now if $s \in [a_k, a_k + \tau_k 2^k/2]$ then $\sqrt{|s - d_k|} \geq \sqrt{\tau_k 2^k/2}$, hence we obtain from the last inequality and from (6.44) that

$$\begin{aligned}
&\mathrm{Bal}\Big(\mu_{K_n}\big|_{[a_k, a_k + \tau_k 2^k]}, [1, b_{k-1}] \cup [d_k, b]\Big)([3 \cdot 2^{k-2}, b_{k-1}]) \\
&= \int_{a_k}^{a_k + \tau_k 2^k/2} \mathrm{Bal}\Big(\delta_s, [1, b_{k-1}] \cup [d_k, b]\Big)([3 \cdot 2^{k-2}, b_{k-1}])\, d\mu_{K_n}(s) \\
&\geq \frac{\sqrt{\tau_k 2^k/2}}{30\pi\sqrt{\theta_k 2^k}} \frac{M_{n,k}\theta_k\sqrt{2^k}}{4 \cdot 10^5\sqrt{2^n}} \geq \frac{M_{n,k}\sqrt{\theta_k}\sqrt{\tau_k 2^k}}{10^8\sqrt{2^n}}
\end{aligned}$$

If we choose

$$\tau_k = \frac{M_{n,0}}{M_{n,k}} \frac{1}{\sqrt{2^k}}, \tag{6.50}$$

then the preceding estimate gives

$$\mathrm{Bal}\Big(\mu_{K_n}\big|_{[a_k, a_k + \tau_k 2^k]}, [1, b_{k-1}] \cup [d_k, b]\Big)([3 \cdot 2^{k-2}, b_{k-1}]) \geq \frac{M_{n,0}}{10^8\sqrt{2^n}} \tag{6.51}$$

because $\tau_k \leq \theta_k$. The choice of τ_k in (6.50) does not contradict the assumed inequality $\tau_k \leq \theta_k$ because for $k > 16$ (for which indices we may have $\theta_k > 0$ at all) the inequality (6.30) gives

$$\tau_k = \frac{M_{n,0}}{M_{n,k}} \frac{1}{\sqrt{2^k}} \leq \left(\frac{17}{16}\right)^k \frac{1}{\sqrt{2^k}} \leq \frac{1}{k} \leq \theta_k, \tag{6.52}$$

where we used our basic assumption that if $\theta_k > 0$ then θ_k is at least as large as $1/k$.

By (6.39)

$$\mu_{K_n}([a_k, a_k + \tau_k 2^k]) \leq \frac{32eM_{n,k}}{\sqrt{2^n}} \sqrt{\theta_k}\sqrt{\tau_k 2^k}, \tag{6.53}$$

hence we get from (6.48) for $t \in [1, 3 \cdot 2^{k-2}]$

$$\begin{aligned}
\frac{d\mathrm{Bal}\Big(\mu_{K_n}\big|_{[a_k, a_k + \tau_k 2^k]}, [1, b_{k-1}] \cup [d_k, b]\Big)(t)}{dt} &\leq 50\frac{\theta_k \tau_k 2^k}{\sqrt{|t - 1|2^k}} \frac{32eM_{n,k}}{\sqrt{2^n}} \\
&\leq 5000\frac{M_{n,0}\theta_k}{\sqrt{|t - 1|}\sqrt{2^n}\sqrt{2^k}}
\end{aligned} \tag{6.54}$$

by the choice of τ_k in (6.50).

The right hand side here is less than

$$5000\pi\frac{M_{n,0}}{\sqrt{2^n}}\theta_k \omega_{[0, b_{k-1}]}(t),$$

and the balayage of $\mu_{[0, b_{k-1}]}$ onto $K_k \subseteq [0, b_{k-1}]$ is just the equilibrium measure of K_k, and hence the balayage of $\mu_{[0, b_{k-1}]}$ onto any set containing K_k is bounded on K_k by the equilibrium measure of K_k. Putting all these together, we obtain that if F is any set containing $K_k \cup [d_k, b_k]$, then for $t \in K_k$, $t \leq 3 \cdot 2^{k-2}$

$$\frac{d\mathrm{Bal}\Big(\mu_{K_n}\big|_{[a_k, a_k + \tau_k 2^k]}, F\Big)(t)}{dt} \leq 5000\pi\frac{M_{n,0}}{\sqrt{2^n}}\theta_k \omega_{K_k}(t).$$

Let now $j \le k-1$ and $t \in [a_j, a_j + d_j]$. It follows from (6.39) applied with k and j (instead of n and k) and then from (6.40) applied with n and j that

$$\omega_{K_k}(t) \le \frac{16eM_{k,j}}{\sqrt{2^k}} \frac{\sqrt{\theta_j}}{\sqrt{|t-a_j|}} \le \frac{16eM_{k,j}}{\sqrt{2^k}} \frac{4 \cdot 10^4 \sqrt{2^n}}{M_{n,j}} \omega_{K_n}(t),$$

therefore for such t we get

$$\begin{aligned} \frac{d\mathrm{Bal}\Big(\mu_{K_n}\big|_{[a_k, a_k+\tau_k 2^k]}, F\Big)(t)}{dt} &\le 5000\pi 16e4 \cdot 10^4 \theta_k \frac{M_{n,0}M_{k,j}}{M_{n,j}} \frac{1}{\sqrt{2^k}} \omega_{K_n}(t) \\ &\le 10^{10} \frac{M_{n,0}M_{k,j}}{M_{n,j}} \frac{1}{\sqrt{2^k}} \omega_{K_n}(t). \end{aligned}$$

Here in view of (6.30)

$$\frac{M_{n,0}M_{k,j}}{M_{n,j}} = \frac{M_{n,0}}{M_{n,j}} \frac{M_{k,j}}{M_{k,k-2}} \le \left(\frac{17}{16}\right)^j \left(\frac{17}{16}\right)^{k-j} = \left(\frac{17}{16}\right)^k,$$

hence

$$\frac{d\mathrm{Bal}\Big(\mu_{K_n}\big|_{[a_k, a_k+\tau_k 2^k]}, F\Big)(t)}{dt} \le 10^{10} \left(\frac{17}{16}\right)^k \frac{1}{\sqrt{2^k}} \omega_{K_n}(t). \qquad (6.55)$$

Using the inequalities (6.38) similar reasoning gives that the same estimate is also true for all $t \in I_j$ provided $0 \le j < k-1$. When $j = k-1$ the same inequality is still true on the interval $[2^{k-1}, 3 \cdot 2^{k-2}]$ – but not necessarily on the right half of I_{k-1}, where the left hand side may be much bigger when t is close to b_{k-1}, c.f. (6.49).

The equilibrium measure for subsets of K_n

Now let

$$I_j^* = [d_j, b_j] = [2^j + \tau_j 2^j, 2^{j+1} - \theta_{j+1} 2^{j+1}]$$

and set (see Figure 17)

$$K_n^* = \cup_{j=0}^{n-1} I_j^*, \qquad (6.56)$$

$$K_{n,k}^* = \cup_{j=0}^{k-1} I_j \bigcup \cup_{j=k}^{n-1} I_j^*. \qquad (6.57)$$

Then $K_{n,n}^* = K_n$, and since $I_0^* = I_0$, we also have $K_{n,1}^* = K_n^*$. Thus, the sets $K_{n,k}^*$ connect in some way the sets K_n and K_n^*.

Let $L_{n,k}$ be the smallest constant for which the estimate

$$\omega_{K_{n,k}^*}(t) \le L_{n,k} \omega_{K_n}(t) \qquad (6.58)$$

is true for all $t \in K_n$, $1 \le t \le 3 \cdot 2^{k-2}$. Since

$$\mu_{K_{n,k}^*} = \mu_{K_{n,k+1}^*} + \mathrm{Bal}\Big(\mu_{K_{n,k+1}^*}\big|_{[a_k, d_k]}, K_{n,k}^*\Big),$$

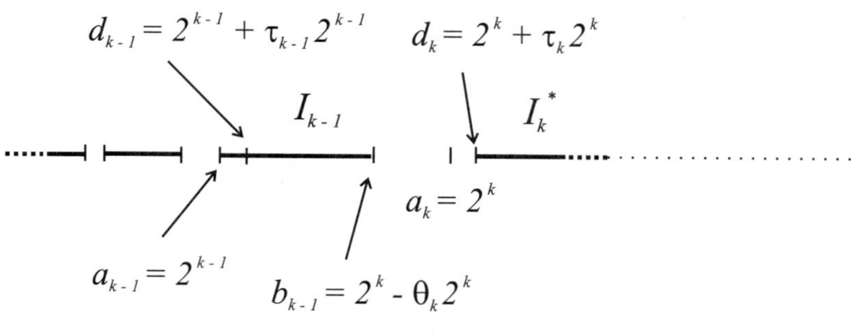

Figure 17:

and here $K^*_{n,k}$ is certainly a set that contains $K_k \cup [d_k, b_k]$, so the inequalities from the end of the previous subsection hold for $F = K^*_{n,k}$. Thus, we obtain from (6.55)

$$L_{n,k} \leq L_{n,k+1} + L_{n,k+1}10^{10}\left(\frac{17}{16}\right)^k \frac{1}{\sqrt{2^k}},$$

i.e.

$$L_{n,k} \leq L_{n,k+1}\left(1 + 10^{10}\left(\frac{17}{16}\right)^k \frac{1}{\sqrt{2^k}}\right) \leq L_{n,k+1}\left(1 + 10^{10}0.8^k\right).$$

On iterating this we finally obtain from $L_{n,n} = 1$ the estimate

$$L_{n,k} \leq \exp(5 \cdot 10^{10}) =: C_0 \tag{6.59}$$

independently of n and k.

In particular, on the interval $[1, 2]$ we have

$$\omega_{K_n^*}(t) \leq C_0 \omega_{K_n}(t). \tag{6.60}$$

Since $\mu_{K_n^*}$ is obtained from $\mu_{K^*_{n,k+2}}$ by taking the balayage of the latter measure onto K_n^*, which amounts to adding to $\mu_{K^*_{n,k+2}}$ the balayage of the restriction of $\mu_{K^*_{n,k+2}}$ to the sets $[a_j, d_j]$, $j = 1, \ldots k+1$ (note that here for $j = 1, \ldots 16$ these intervals are certainly empty), we can infer

$$\begin{aligned}
\mu_{K_n^*}(I_k^*) &\leq \mu_{K^*_{n,k+2}}(I_k^*) + \sum_{j=1}^{k+1} \mu_{K^*_{n,k+2}}([a_k, d_k]) \\
&\leq L_{n,k+2}\mu_{K_n}(I_k^*) + \sum_{j=1}^{k+1} L_{n,k+2}\mu_{K_n}([a_k, d_k]) \\
&\leq C_0\mu_{K_n}(I_k) + C_0\sum_{j=1}^{k+1} \mu_{K_n}([a_k, d_k]).
\end{aligned}$$

The first term on the right was estimated in (6.45), while the terms in the sum were estimated in (6.43) (recall that $d_k = a_k + \tau_k 2^k \leq a_k + \theta_k 2^k$), and so we obtain

$$\mu_{K_n^*}(I_k^*) \leq C_0 100\frac{M_{n,k}\sqrt{2^k}}{\sqrt{2^n}} + \sum_{j=1}^{k+1} C_0 32e\frac{M_{n,j}\theta_j\sqrt{2^j}}{\sqrt{2^n}}.$$

Here $\theta_j < 1/16$, and in view of (6.30) we have

$$\begin{aligned}\sum_{j=1}^{k+1} \frac{M_{n,j}\sqrt{2^j}}{\sqrt{2^n}} &\leq \frac{M_{n,k}\sqrt{2^k}}{\sqrt{2^n}} \sum_{j=1}^{k+1} \left(\frac{17}{16}\right)^{|k-j|} \frac{1}{(\sqrt{2})^{k-j}} \\ &\leq 10\frac{M_{n,k}\sqrt{2^k}}{\sqrt{2^n}},\end{aligned}$$

and so we obtain

$$\mu_{K_n^*}(I_k^*) \leq 200C_0\frac{M_{n,k}\sqrt{2^k}}{\sqrt{2^n}}. \tag{6.61}$$

Completely similar calculation gives that if $K_n^* \subseteq E_n \subseteq K_n$ is any compact set in between K_n^* and K_n, then

$$\mu_{E_n}([2^k, 2^{k+1}]) \leq 200C_0\frac{M_{n,k}\sqrt{2^k}}{\sqrt{2^n}}. \tag{6.62}$$

Estimation of Green functions

After all these, we are in the position to estimate $g_{\mathbf{C}\setminus K_n}(0)$, i.e. the value at the origin of the Green function for the complement of K_n.

Since $1 \in K_n$, we obtain from (2.18)

$$g_{\mathbf{C}\setminus K_n}(0) = g_{\mathbf{C}\setminus K_n}(0) - g_{\mathbf{C}\setminus K_n}(1) = \int_{K_n} \log\frac{|t|}{|t-1|}d\mu_{K_n}(t).$$

Let us break the integral as

$$\int_1^2 + \sum_{j=1}^{n-1}\int_{I_j}.$$

Here by (6.41)

$$\int_1^2 \log\frac{|t|}{|t-1|}d\mu_{K_n}(t) \leq 16e\frac{M_{n,0}}{\sqrt{2^n}}\int_1^2 \log\frac{t}{t-1}\frac{1}{\sqrt{t-1}dt} \leq 500\frac{M_{n,0}}{\sqrt{2^n}},$$

and

$$\int_1^2 \log\frac{|t|}{|t-1|}d\mu_{K_n}(t) \geq \frac{1}{16e}\frac{M_{n,0}}{\sqrt{2^n}}\int_1^2 \log\frac{t}{t-1}\frac{1}{\sqrt{t-1}}dt \geq \frac{1}{100}\frac{M_{n,0}}{\sqrt{2^n}}.$$

On the other hand, (6.45) gives

$$\int_{I_k} \log\frac{|t|}{|t-1|}d\mu_{K_n}(t) \leq \log\frac{2^k}{2^k-1}100\frac{M_{n,k}\sqrt{2^k}}{\sqrt{2^n}} \leq 200\frac{M_{n,k}}{\sqrt{2^k}\sqrt{2^n}},$$

and so we obtain from (6.30)

$$\sum_{k=1}^{n-1}\int_{I_k} \log\frac{|t|}{|t-1|}d\mu_{K_n}(t) \leq 200\frac{M_{n,0}}{\sqrt{2^n}}\sum_{k=1}^{n-1}\left(\frac{17}{16}\right)^k\frac{1}{(\sqrt{2})^k} \leq 900\frac{M_{n,0}}{\sqrt{2^n}}.$$

In summary, we have

$$\frac{1}{100}\frac{M_{n,0}}{\sqrt{2^n}} \le g_{K_n}(0) \le 1000\frac{M_{n,0}}{\sqrt{2^n}}. \tag{6.63}$$

Consider now the set K_n^* from (6.56) and the Green function of its complement. Since $K_n^* \subset K_n$, we have

$$g_{\mathbf{C}\setminus K_n^*}(0) \ge g_{\mathbf{C}\setminus K_n}(0) \ge \frac{1}{100}\frac{M_{n,0}}{\sqrt{2^n}}. \tag{6.64}$$

If we use (6.60) and use (6.61) instead of the upper estimate in (6.45), then the computation that we have just carried out gives that

$$g_{\mathbf{C}\setminus K_n^*}(0) \le (100C_0 + 2C_0 \cdot 900)\frac{M_{n,0}}{\sqrt{2^n}} < 2000C_0\frac{M_{n,0}}{\sqrt{2^n}}, \tag{6.65}$$

where C_0 is the absolute constant from (6.59).

Finally let us state the immediate corollary of (6.63) – (6.65) that if $K_n^* \subseteq E_n \subseteq K_n$ is any compact set, then

$$\frac{1}{100}\frac{M_{n,0}}{\sqrt{2^n}} \le g_{\mathbf{C}\setminus E_n}(0) \le 2000C_0\frac{M_{n,0}}{\sqrt{2^n}}. \tag{6.66}$$

Moving endpoints of sets

Our ultimate aim is to show that by appropriately moving the left endpoints of the intervals $[a_k, b_k]$ by an amount that does not exceed $\tau_k 2^k$, then we arrive at a set E_n for which an appropriate constant multiple of the equilibrium measure has integer masses on each subinterval of E_n.

To this end let $\underline{\mathbf{x}} = (x_0, \dots, x_{n-1})$ be a point in $[0,1]^n$, and set

$$E_{\underline{\mathbf{x}}} = \bigcup_{k=0}^{n-1}[a_k + x_k\tau_k 2^k, b_k] = \bigcup_{k=0}^{n-1}[2^k + x_k\tau_k 2^k, 2^{k+1} - \theta_{k+1}2^{k+1}]. \tag{6.67}$$

Our set E_n will be one of the $E_{\underline{\mathbf{x}}}$'s for appropriate $\underline{\mathbf{x}} = \underline{\mathbf{x}}_n \in [0,1]^n$. Note that if $\theta_k = 0$ then $\tau_k = 0$ and then $E_{\underline{\mathbf{x}}}$ does not depend on the variable x_k.

Let also consider

$$F_{k,\underline{\mathbf{x}}} = \bigcup_{j=0}^{k-1}[a_j + x_j\tau_j 2^j, b_j], \qquad k = 1, \dots, n \tag{6.68}$$

and

$$f_k(\underline{\mathbf{x}}) = \mu_{E_{\underline{\mathbf{x}}}}(F_{k,\underline{\mathbf{x}}})\frac{\sqrt{2^n}}{M_{n,0}}T, \tag{6.69}$$

where

$$10^9 < T < 10^{10} \tag{6.70}$$

is a number such that
$$\frac{\sqrt{2^n}}{M_{n,0}}T$$
is an integer. In view of (6.30) $M_{n,0}$ is much smaller than $\sqrt{2^n}$, so there is such a T satisfying (6.70). Then
$$f_n(\underline{\mathbf{x}}) \equiv \frac{\sqrt{2^n}}{M_{n,0}}T$$
is an integer, and we show that for an appropriate choice of $\underline{\mathbf{x}} \in [0,1]^n$ all the functions $f_k(\underline{\mathbf{x}})$ with $\theta_k > 0$, $1 \le k \le n$ are integers. Let us claim this as

Proposition 6.6 *There is an $\underline{\mathbf{x}} \in [0,1]^n$ such that for all $k = 1, \ldots, n$ for which $\theta_k > 0$, the value $f_k(\underline{\mathbf{x}})$, is an integer.*

This proposition implies

Corollary 6.7 *If $\underline{\mathbf{x}}$ is the vector from Proposition* 6.6*, and if J is a subinterval of $E_{\underline{\mathbf{x}}}$, then*
$$\mu_{E_{\underline{\mathbf{x}}}}(J)\frac{\sqrt{2^n}}{M_{n,0}}T$$
is an integer.

Proof. In fact, J can be written as $F_{k+1,\underline{\mathbf{x}}} \setminus F_{l,\underline{\mathbf{x}}}$ where $a_l + x_l\tau_l 2^l$ is the left and b_k is the right endpoint of J, and then here $\theta_l > 0$, $\theta_{k+1} > 0$, but for all other index $l < j \le k$ we have $\theta_j = 0$. Now the corollary immediately follows from the proposition.

■

We continue with the proof of Proposition 6.6. As we have just seen, we do not have to worry about f_n – the choice of T takes care of that.

The proof will follow from Brouwer's fixed point theorem. Let us recall that the θ in (6.24) has been arbitrary up to now, and all the estimates so far depended only on the fact that $\theta_j < 1/16$ and that if $\theta_j > 0$, then $\theta_j > 1/j$. Now we take into account the properties (6.24) with some small $\theta < 1/16$. With the help of (6.24) we verify

Proposition 6.8 *There is an absolute constant $\theta > 0$ such that if $\theta_k > 0$, $1 \le k \le n-1$, and $\underline{\mathbf{x}}, \underline{\mathbf{y}} \in [0,1]^n$ are any two points with $(\underline{\mathbf{x}})_k = 1$ and $(\underline{\mathbf{y}})_k = 0$, then*
$$f_k(\underline{\mathbf{x}}) - f_k(\underline{\mathbf{y}}) > 8. \tag{6.71}$$
Note again that if $\theta_j = 0$ then the functions $f_k(\underline{\mathbf{x}})$ do not depend on the variable x_j.

First we show that Proposition 6.6 follows from Proposition 6.8.

Proof of Proposition 6.6. Let θ be as in Proposition 6.8. Then (6.71) is true, hence for any k with $\theta_k > 0$ there is an integer z_k with the property
$$\sup_{(\underline{\mathbf{y}})_k=0} f_k(\underline{\mathbf{y}}) + 2 < z_k < \inf_{(\underline{\mathbf{x}})_k=1} f_k(\underline{\mathbf{x}}) - 2. \tag{6.72}$$

We claim that there is an $\underline{\mathbf{x}} \in [0,1]^n$ such that $f_k(\underline{\mathbf{x}}) = z_k$ for all $k = 1, \dots, n-1$, for which $\theta_k > 0$, and that proves Proposition 6.6.

In fact, let $k_1, \dots, k_m$ be those $k \in \{1, \dots, n-1\}$ for which $\theta_k > 0$, and consider the map

$$\mathbf{F}(\underline{\mathbf{x}}) = (z_{k_1} - f_{k_1}(\underline{\mathbf{x}}) + 1/2, \dots, z_{k_m} - f_{k_m}(\underline{\mathbf{x}}) + 1/2).$$

We have to show that $\mathbf{F}(\underline{\mathbf{x}}) = \mathbf{1/2}$ for some $\underline{\mathbf{x}} \in [0,1]^n$, where

$$\mathbf{1/2} = (1/2, \dots, 1/2) \in [0,1]^m.$$

We have already mentioned that if $\theta_j = 0$, then the function $\mathbf{F}(\underline{\mathbf{x}})$ does not depend on x_j, therefore we may write $\mathbf{F}(\underline{\mathbf{x}})$ as $\mathbf{F}(\underline{\mathbf{X}})$, where $\underline{\mathbf{X}} = (x_{k_1}, \dots, x_{k_m})$. If $(\underline{\mathbf{X}})_j = 0$, then (6.72) shows that $\mathbf{F}(\underline{\mathbf{X}})$ lies in the half space $y_j > 1$ of $\mathbf{R}^m$, but if $(\underline{\mathbf{X}})_j = 1$, then $\mathbf{F}(\underline{\mathbf{X}})$ lies in the half space $y_j < 0$. Now suppose that there is no $\underline{\mathbf{X}}$ with $\mathbf{F}(\underline{\mathbf{X}}) = \mathbf{1/2}$. Then we can consider for all $\underline{\mathbf{X}} \in [0,1]^m$ the half line starting from $\mathbf{1/2}$ and going through $\mathbf{F}(\underline{\mathbf{X}})$, and let $P(\underline{\mathbf{X}})$ be a point where this half line intersects the boundary of $[0,1]^m$. It is clear that the mapping $\underline{\mathbf{X}} \to P(\underline{\mathbf{X}})$ is a continuous mapping of $[0,1]^m$ into itself, and what we have just said implies that this mapping does not have a fixed point. In fact, since it is a mapping onto the boundary, an inner point of $[0,1]^m$ cannot be a fixed point. But if $\underline{\mathbf{X}}$ lies on the boundary of $[0,1]^m$, then for some j we have either $(\underline{\mathbf{X}})_j = 0$ or $(\underline{\mathbf{X}})_j = 1$. In the former case $\mathbf{F}(\underline{\mathbf{X}})$ lies in the half space $y_j > 1$, so the intersection of the half line from $\mathbf{1/2}$ through $\mathbf{F}(\underline{\mathbf{X}})$ cannot intersect the boundary in a point $\underline{\mathbf{y}}$ with $(\underline{\mathbf{y}})_j = 0$, in particular, $P(\underline{\mathbf{X}}) \neq \underline{\mathbf{X}}$. In a similar fashion do we get that if $(\underline{\mathbf{X}})_j = 1$ then $P(\underline{\mathbf{X}}) \neq \underline{\mathbf{X}}$, so, indeed, this mapping does not have a fixed point.

But this contradicts the Brouwer fixed point theorem, and this contradiction shows that there must be an $\underline{\mathbf{X}}$ with $\mathbf{F}(\underline{\mathbf{X}}) = \mathbf{1/2}$ as we claimed above.

■

Proof of Proposition 6.8

First we establish some monotonicity properties of the functions f_k. If $j \geq k$ and $\theta_j > 0$ (otherwise $f_k(\underline{\mathbf{x}})$ does not depend on x_j), then for $x_j < x'_j$ and for

$$\underline{\mathbf{x}}' = (x_0, \dots, x_{j-1}, x'_j, x_{j+1}, \dots, x_{n-1})$$

the set $E_{\underline{\mathbf{x}}'}$ is a subset of $E_{\underline{\mathbf{x}}}$, hence its equilibrium measure is obtained by taking balayage onto this set of the equilibrium measure of $E_{\underline{\mathbf{x}}}$, i.e.

$$\mu_{E_{\underline{\mathbf{x}}'}} = \mu_{E_{\underline{\mathbf{x}}}}\Big|_{E_{\underline{\mathbf{x}}'}} + \mathrm{Bal}\Big(\mu_{E_{\underline{\mathbf{x}}}}\Big|_{[a_j + x_j\tau_j 2^j,\, a_j + x'_j\tau_j 2^j]}, E_{\underline{\mathbf{x}}'}\Big).$$

This gives

$$\frac{M_{n,0}}{\sqrt{2^n T}}\,(f_k(\underline{\mathbf{x}}) - f_k(\underline{\mathbf{x}}')) = \mathrm{Bal}\Big(\mu_{E_{\underline{\mathbf{x}}}}\Big|_{[a_j + x_j\tau_j 2^j,\, a_j + x'_j\tau_j 2^j]}, E_{\underline{\mathbf{x}}'}\Big)(F_{k,\underline{\mathbf{x}}'}).$$

In particular, $f_k(\underline{\mathbf{x}})$ is an increasing function of x_j. Similar consideration shows that for $j \le k-1$ the function $f_k(\underline{\mathbf{x}})$ is a decreasing function of x_j.

We claim that there is a $\theta > 0$ such that if (6.24) is satisfied and $\theta_k > 0$, then

$$f_k(0,\dots,0,\overbrace{0}^{k+1},1,1,\dots,1) - f_k(0,\dots,0) < 1 \tag{6.73}$$

and

$$f_k(0,\dots,0,\overbrace{1}^{k+1},0,0,\dots,0) - f_k(1,\dots,1,1,\overbrace{1}^{k+1},0,0,\dots,0) < 1. \tag{6.74}$$

First we show that these already prove Proposition 6.8. Let $x_i, y_i \in [0,1]$ be arbitrary. We get from the monotonicity properties of f_k

$$\begin{aligned} f_k(x_0,\dots,x_{k-1},0,x_{k+1},\dots,x_{n-1}) &\le f_k(0,\dots,0,\overbrace{0}^{k+1},1,1,\dots,1) \\ &< f_k(0,\dots,0)+1 \end{aligned} \tag{6.75}$$

where in the last step we used (6.73). Let $\underline{\mathbf{z}}_k = (0,\dots,0,\overbrace{1}^{k+1},0,\dots,0)$. We have assumed that k is an integer for which $\theta_k > 0$, so by the definition of the functions $f_k(\underline{\mathbf{x}})$ we get

$$\begin{aligned} f_k(\underline{\mathbf{z}}_k) - f_k(\underline{\mathbf{0}}) &= \mathrm{Bal}\Big(\mu_{K_n}\big|_{[a_k,\, a_k+\tau_k 2^k]}, E_{\underline{\mathbf{z}}_k}\Big)(K_k)\frac{\sqrt{2^n}}{M_{n,0}}T \\ &\ge \mathrm{Bal}\Big(\mu_{K_n}\big|_{[a_k,\, a_k+\tau_k 2^k]}, E_{\underline{\mathbf{z}}_k}\Big)([a_{k-1},b_{k-1}])\frac{\sqrt{2^n}}{M_{n,0}}T. \end{aligned}$$

Here the balayage can be obtained in two steps: first take the balayage onto $[1,b_{k-1}] \cup [d_k,b_{n-1}]$, and then take the balayage of that measure onto $E_{\underline{\mathbf{z}}_k}$. Thus, we obtain from (6.51) with $b = b_{n-1}$ that

$$f_k(\underline{\mathbf{z}}_k) - f_k(\underline{\mathbf{0}}) \ge \frac{M_{n,0}}{10^8\sqrt{2^n}}\frac{\sqrt{2^n}}{M_{n,0}}T > 10. \tag{6.76}$$

by the choice of T in (6.70).

Hence we can continue (6.75) as

$$\begin{aligned} f_k(0,\dots,0)+1 &< f_k(0,\dots,0,\overbrace{1}^{k+1},0,\dots,0) - 9 \\ &\le f_k(1,\dots,1,\overbrace{1}^{k+1},0,\dots,0) - 8 \\ &\le f_k(y_0,\dots,y_{k-1},1,y_{k+1},\dots,y_{n-1}) - 8, \end{aligned}$$

where in the last but one inequality we used (6.74) and in the last one the monotonicity properties of f_k. This with (6.75) proves Proposition 6.8 pending the proofs of (6.73) and (6.74).

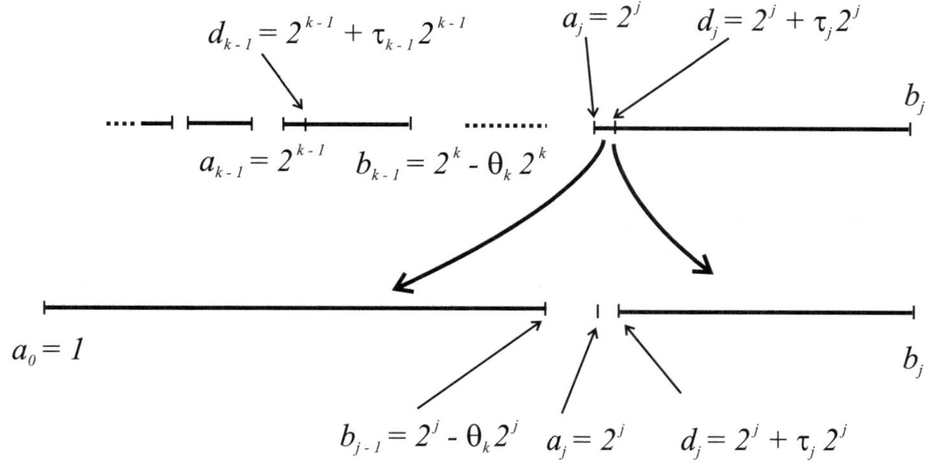

Figure 18:

Proof of (6.73)

In what follows we shall repeatedly use that if $C \subset A \subset B$, then on C the balayage of any measure onto A is at least as large as the balayage onto B (this follows from the fact that taking balayage onto A can be done first by taking balayage onto B and then onto A).

For

$$\underline{\mathbf{x}} = (0, \ldots, 0, \overbrace{0}^{k+1}, 1, 1, \ldots, 1)$$

the set $E_{\underline{\mathbf{x}}}$ is nothing else than $K^*_{n,k+1}$ (see (6.57)) and $E_{\underline{\mathbf{0}}}$ is just K_n (see (6.67)), furthermore if $x_j = 0$ for $0 \le j < k$ then we have $F_{k,\underline{\mathbf{x}}} = K_k$ (see (6.68)). Therefore, (see (6.69))

$$\begin{aligned}
&\frac{M_{n,0}}{\sqrt{2^n T}}\Big(f_k(0, \ldots, 0, \overbrace{0}^{k+1}, 1, 1, \ldots, 1) - f_k(0, \ldots, 0, 0, \overbrace{0}^{k+1}, 0, 0, \ldots, 0)\Big) \\
&= \sum_{j=k+1}^{n-1} \mathrm{Bal}\Big(\mu_{K_n}\big|_{[a_j,\, d_j]}, K^*_{n,k+1}\Big)(K_k) \\
&\le \sum_{j=k+1}^{n-1} \mathrm{Bal}\Big(\mu_{K_n}\big|_{[a_j,\, d_j]}, K^*_{j+1,k+1}\Big)(K_k). \qquad (6.77)
\end{aligned}$$

Let (see Figure 18)

$$\nu_j = \mathrm{Bal}\Big(\mu_{K_n}\big|_{[a_j,\,d_j]},\,[1,b_{j-1}]\cup[d_j,b_j]\Big).$$

By (6.54) the density of ν_j on the interval $[1, 3\cdot 2^{j-2}]$ satisfies

$$\frac{d\nu_j(t)}{dt} \le 5000\frac{M_{n,0}\theta_j}{\sqrt{|t-1|}\sqrt{2^n}\sqrt{2^j}}, \tag{6.78}$$

and (see Figure 19, where the arrows indicate sweeping out – balayaging – masses, the dotted arrows indicate that the measure is increased if we do not take balayage onto those parts, and the second and third parts of the picture schematically show how the balayage measures in question are taken)

$$\begin{aligned}
&\mathrm{Bal}\Big(\mu_{K_n}\big|_{[a_j,\,d_j]},\,K^*_{j+1,k+1}\Big)(K_k)\\
&= \mathrm{Bal}\Big(\mu_{K_n}\big|_{[a_j,\,d_j]},\,K^*_{j+1,k+1}\Big)([1,2^k])\\
&= \nu_j([1,2^k]) + \sum_{s=k+1}^{j-1}\mathrm{Bal}\Big(\nu_j\big|_{[b_{s-1},\,d_s]},\,K^*_{j+1,k+1}\Big)([1,2^k]),
\end{aligned}$$

and here the last sum is increased if in the s-th term instead of $K^*_{j+1,k+1}$, we take the balayage onto $K^*_{s+1,k+1}$, i.e.

$$\begin{aligned}
\mathrm{Bal}\Big(\mu_{K_n}\big|_{[a_j,\,d_j]},\,K^*_{j+1,k+1}\Big)([1,2^k]) &\le \nu_j([1,2^k])\\
&+ \sum_{s=k+1}^{j-1}\mathrm{Bal}\Big(\nu_j\big|_{[b_{s-1},\,d_s]},\,K^*_{s+1,k+1}\Big)([1,2^k]).
\end{aligned} \tag{6.79}$$

Here the first term on the right can be estimated by integrating (6.78) from 1 to 2^k:

$$\nu_j([1,2^k]) \le 5000\frac{M_{n,0}}{\sqrt{2^n}}\frac{\theta_j}{\sqrt{2^j}}2\sqrt{2^k} = 10^4\frac{M_{n,0}}{\sqrt{2^n}}\frac{\theta_j}{\sqrt{2^j}}\sqrt{2^k}. \tag{6.80}$$

Let now

$$\nu_{j,s} = \mathrm{Bal}\Big(\nu_j\big|_{[b_{s-1},\,d_s]},\,[1,b_{s-1}]\cup[d_s,b_s]\Big).$$

It follows from (6.47) that for $t\in[1,3\cdot 2^{s-2}]$

$$\frac{d\nu_{j,s}(t)}{dt} \le 50\sqrt{2}\frac{\theta_s}{\sqrt{|t-1|}\sqrt{2^s}}\nu_j([b_{s-1},d_s]),$$

which in view of (6.78) can be estimated as

$$\frac{d\nu_{j,s}(t)}{dt} \le 50\sqrt{2}\frac{\theta_s}{\sqrt{|t-1|}\sqrt{2^s}}5000\frac{M_{n,0}\theta_j}{\sqrt{2^{s-1}}\sqrt{2^n}\sqrt{2^j}}2\theta_s 2^s = 10^6\frac{M_{n,0}}{\sqrt{2^n}}\frac{\theta_j\theta_s^2}{\sqrt{t-1}\sqrt{2^j}}. \tag{6.81}$$

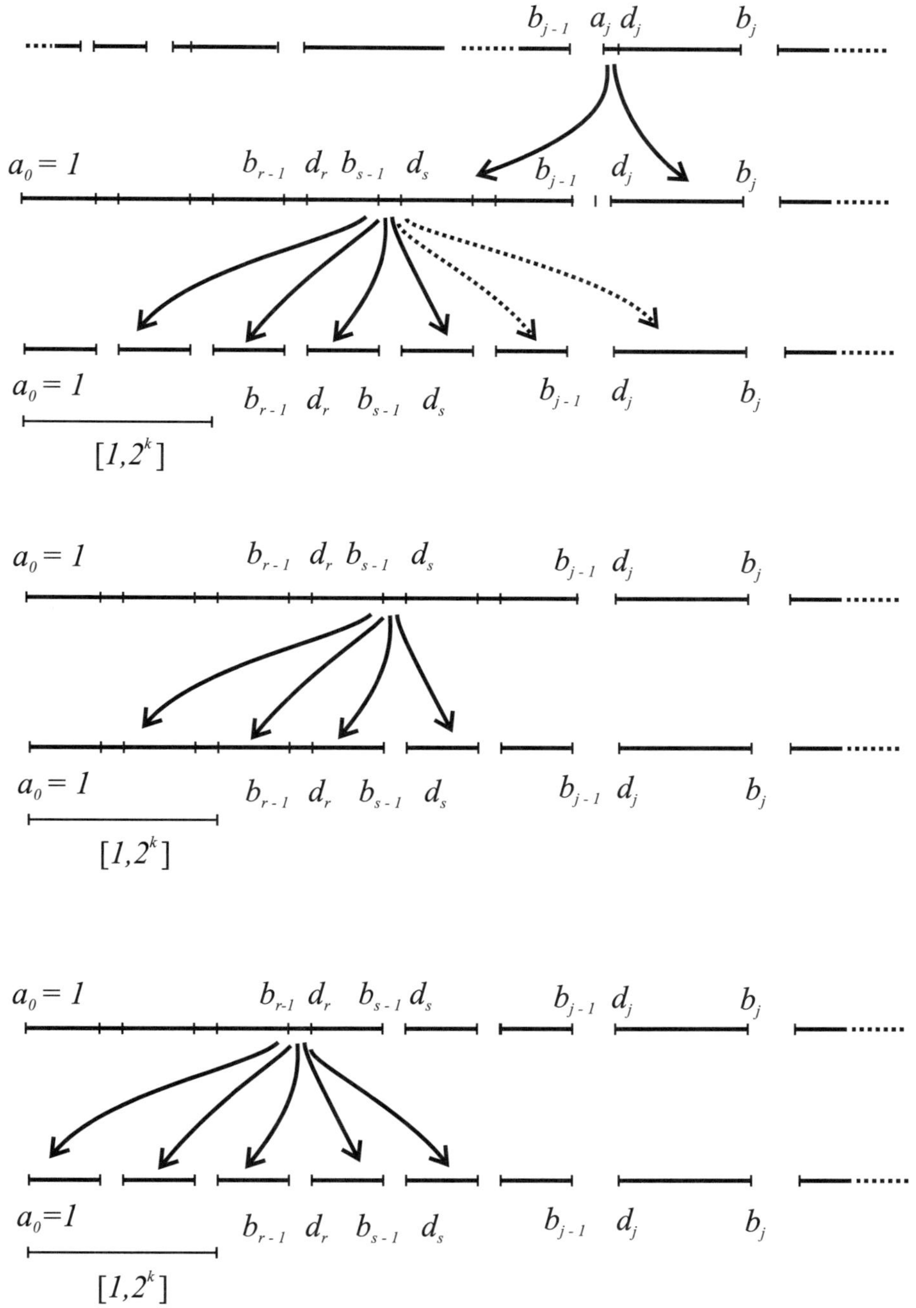

b_{j-1} a_j d_j b_j
$a_0 = 1$ b_{r-1} d_r b_{s-1} d_s b_{j-1} d_j b_j
$a_0 = 1$ b_{r-1} d_r b_{s-1} d_s b_{j-1} d_j b_j
$[1,2^k]$
$a_0 = 1$ b_{r-1} d_r b_{s-1} d_s b_{j-1} d_j b_j
$a_0 = 1$ b_{r-1} d_r b_{s-1} d_s b_{j-1} d_j b_j
$[1,2^k]$
$a_0 = 1$ b_{r-1} d_r b_{s-1} d_s b_{j-1} d_j b_j
$a_0=1$ b_{r-1} d_r b_{s-1} d_s b_{j-1} d_j b_j
$[1,2^k]$

Figure 19:

Using this we can write for the terms in the sum (6.79)

$$\begin{aligned}
&\mathrm{Bal}\Big(\nu_j\Big|_{[b_{s-1},\,d_s]},K^*_{s+1,k+1}\Big)([1,2^k]) \qquad (6.82)\\
&= \nu_{j,s}([1,2^k]) + \sum_{r=k+1}^{s-1}\mathrm{Bal}\Big(\nu_{j,s}\Big|_{[b_r,\,d_r]},K^*_{s+1,r+1}\Big)([1,2^k])\\
&\le \nu_{j,s}([1,2^k]) + \sum_{r=k+1}^{s-1}\nu_{j,s}([b_{r-1},d_r])\\
&\le 10^6\frac{M_{n,0}}{\sqrt{2^n}}\frac{\theta_j\theta_s^2}{\sqrt{2^j}}2\sqrt{2^k} + 10^6\frac{M_{n,0}}{\sqrt{2^n}}\sum_{r=k+1}^{s-1}\frac{\theta_j\theta_s^2}{\sqrt{2^j}}\frac{2\theta_r 2^r}{\sqrt{2^{r-1}}}.
\end{aligned}$$

Summing up, it follows from (6.77), (6.79), (6.80) and (6.82) that

$$\begin{aligned}
&\frac{M_{n,0}}{\sqrt{2^n}T}\Big(f_k(0,\dots,0,\overbrace{0}^{k+1},1,1,\dots,1) - f_k(0,\dots,0,0,\overbrace{0}^{k+1},0,0,\dots,0)\Big)\\
&\le \sum_{j=k+1}^{n-1}10^4\frac{M_{n,0}}{\sqrt{2^n}}\frac{\theta_j}{\sqrt{2^j}}\sqrt{2^k}\\
&+ \sum_{j=k+1}^{n-1}\sum_{s=k+1}^{j-1}10^6 2\frac{M_{n,0}}{\sqrt{2^n}}\frac{\theta_j\theta_s^2}{\sqrt{2^j}}\sqrt{2^k} + \sum_{j=k+1}^{n-1}\sum_{s=k+1}^{j-1}\sum_{r=k+1}^{s-1}10^6 2\sqrt{2}\frac{M_{n,0}}{\sqrt{2^n}}\frac{\theta_j\theta_s^2}{\sqrt{2^j}}\theta_r\sqrt{2^r}\\
&= \frac{M_{n,0}}{\sqrt{2^n}}\Big(\sum_1+\sum_2+\sum_3\Big). \qquad (6.83)
\end{aligned}$$

Here, in view of $\theta_j \le \theta < 1/16$

$$\Sigma_1 \le 10^4\theta\frac{\sqrt{2}}{\sqrt{2}-1} < 4\cdot 10^4\theta,$$

and by rearrangement we get

$$\Sigma_2 \le 2\cdot 10^6\theta^3\sqrt{2^k}\sum_{s=k+1}^{n-2}\sum_{j=s+1}^{n-1}\frac{1}{\sqrt{2^j}} \le 2\cdot 10^6\theta^3\left(\frac{\sqrt{2}}{\sqrt{2}-1}\right)^2 < 2\cdot 10^5\theta.$$

Finally we write the third sum as

$$\begin{aligned}
\Sigma_3 &= 10^6 2\sqrt{2}\sum_{j=k+1}^{n-1}\sum_{s=k+1}^{j-1}\sum_{r=k+1}^{s-1}\frac{\theta_j\theta_s^2}{\sqrt{2^j}}\theta_r\sqrt{2^r}\\
&= 10^6 2\sqrt{2}\sum_{p=1}^{\infty}\sum_{q=1}^{\infty}\sum_{j=k+1}^{n-1}\frac{\theta_j\theta_{j-p}^2\theta_{j-p-q}}{\sqrt{2^j}}\sqrt{2^{j-p-q}}
\end{aligned}$$

with the understanding that if $j-p\le k$ then in the last sum on the right we set $\theta_{j-p}=0$ and if $j-p-q\le k$, then we set $\theta_{j-p-q}=0$. By Hölder's inequality and (6.24)

$$\sum_{j=k+1}^{n-1}\theta_j\theta_{j-p}^2\theta_{j-p-q} \le \theta\sum_j\theta_l^3 \le \theta^2,$$

so we finally obtain

$$\Sigma_3 \leq 10^6 2\sqrt{2}\theta^2 \sum_{p=1}^{\infty}\sum_{q=1}^{\infty} \frac{1}{\sqrt{2^{p+q}}} \leq 10^6 2\sqrt{2}\theta^2 \left(\frac{2}{\sqrt{2}-1}\right)^2 < 3\cdot 10^6.$$

What we have proved since (6.83) gives

$$\begin{aligned} f_k(0,\dots,0,\overbrace{0}^{k+1},1,1,\dots,1) \quad &- \quad f_k(0,\dots,0,0,\overbrace{0}^{k+1},0,0,\dots,0) \\ &< \quad T(4\cdot 10^4 + 2\cdot 10^5 + 3\cdot 10^6)\theta < 10^{17}\theta, \end{aligned}$$

where in the last inequality we used that $T < 10^{10}$ by (6.70). Hence the choice $\theta = 10^{-17}$ is appropriate to achieve (6.73).

∎

Proof of (6.74)

Let $s \in [a_k, d_k] = [2^k, 2^k + \tau_k 2^k]$. By (2.25) we have for $t \in [d_{k-1}, b_{k-1}] \cup [d_k, b_{n-1}]$

$$\frac{d\mathrm{Bal}\Big(\delta_s, [d_{k-1}, b_{k-1}] \cup [d_k, b_{n-1}]\Big)(t)}{dt} = \tilde{H}(t)$$

with

$$\tilde{H}(t) = \tilde{H}_{k,b,s}(t) = \frac{\sqrt{|s-d_{k-1}||s-b_{k-1}||s-d_k||s-b_{n-1}|}}{\pi\sqrt{|t-d_{k-1}||t-b_{k-1}||t-d_k||t-b_{n-1}|}} \frac{|t-\underline{\mathbf{z}}|}{|s-\underline{\mathbf{z}}|} \frac{1}{|t-s|},$$

where $\underline{\mathbf{z}} = \underline{\mathbf{z}}_{k,s}$ is determined by the equation

$$\left(\int_{-\infty}^{d_{k-1}} + \int_{b_{n-1}}^{\infty}\right) \tilde{H}(t)\mathrm{sign}(t-\underline{\mathbf{z}})dt = 0.$$

Here either $\underline{\mathbf{z}} \leq d_{k-1}$ or $\underline{\mathbf{z}} \geq b_{n-1}$, and in either case we get for $t \in [3\cdot 2^{k-1}, 3\cdot 2^{n-2}]$ that

$$\tilde{H}(t) \leq \frac{\sqrt{2^{k+1}(2\theta_k 2^k)(\tau_k 2^k)2^n}}{\pi\sqrt{(t/2)(t/4)(t/4)2^{n-2}}} \frac{t}{2^{k-1}} \frac{1}{t/4} = \frac{64\sqrt{2}}{\pi} \frac{\sqrt{\theta_k \tau_k}}{t^{3/2}} \sqrt{2^k}, \tag{6.84}$$

while for $s \in [b_{k-1}, d_k]$ the same computation shows that we have for $t \in [3\cdot 2^{k-1}, 3\cdot 2^{n-2}]$

$$\tilde{H}(t) \leq \frac{128}{\pi} \frac{\theta_k}{t^{3/2}} \sqrt{2^k}, \tag{6.85}$$

Exactly as in the proof of (6.73) we get that with

$$K^{**}_{n,k+1} = \cup_{l=1}^{k} I_l^* \bigcup \cup_{l=k+1}^{n-1} I_l$$

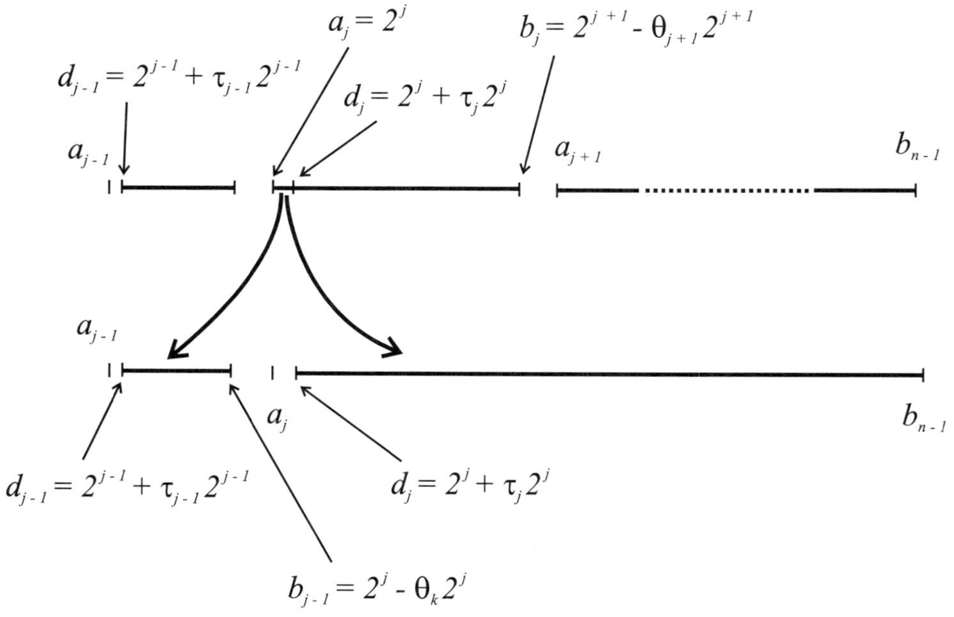

Figure 20:

we have

$$\frac{M_{n,0}}{\sqrt{2^n T}}\Big(f_k(0,\ldots,0,\overbrace{1}^{k+1},0,0,\ldots,0) - f_k(1,\ldots,1,1,\overbrace{1}^{k+1},0,0,\ldots,0)\Big)$$
$$= \sum_{j=1}^{k-1} \mathrm{Bal}\Big(\mu_{(K_n\setminus[a_k,d_k))}\Big|_{[a_j,\,d_j]}, K^{**}_{n,k+1}\Big)(I^*_k \bigcup \cup_{l=k+1}^{n-1} I_l),$$
$$\le \sum_{j=1}^{k-1} \mathrm{Bal}\Big(\mu_{(K_n\setminus[a_k,d_k))}\Big|_{[a_j,\,d_j]}, \cup_{l=j-1}^{k-1} I^*_l \bigcup [b_{k-1}, b_{n-1}]\Big)([b_{k-1}, b_{n-1}]). \tag{6.86}$$

Now let (see Figure 20)

$$\rho_j = \mathrm{Bal}\Big(\mu_{(K_n\setminus[a_k,d_k))}\Big|_{[a_j,\,d_j]}, [d_{j-1}, b_{j-1}] \cup [d_j, b_{n-1}]\Big).$$

Then the j-th term in the last sum is equal to (see Figure 21, where the arrows indicate sweeping out – balayaging – masses, the dotted arrows indicate that the measure is increased if we do not take balayage onto those parts, and the second and third parts of the figure schematically show how the balayage measures in question are taken)

$$\rho_j([b_{k-1}, b_{n-1}]) + \sum_{s=j+1}^{k-1} \mathrm{Bal}\Big(\rho_j\Big|_{[b_{s-1},\,d_s]}, \cup_{l=j-1}^{k-1} I^*_l \cup [b_{k-1}, b_{n-1}]\Big)([b_{k-1}, b_{n-1}])$$

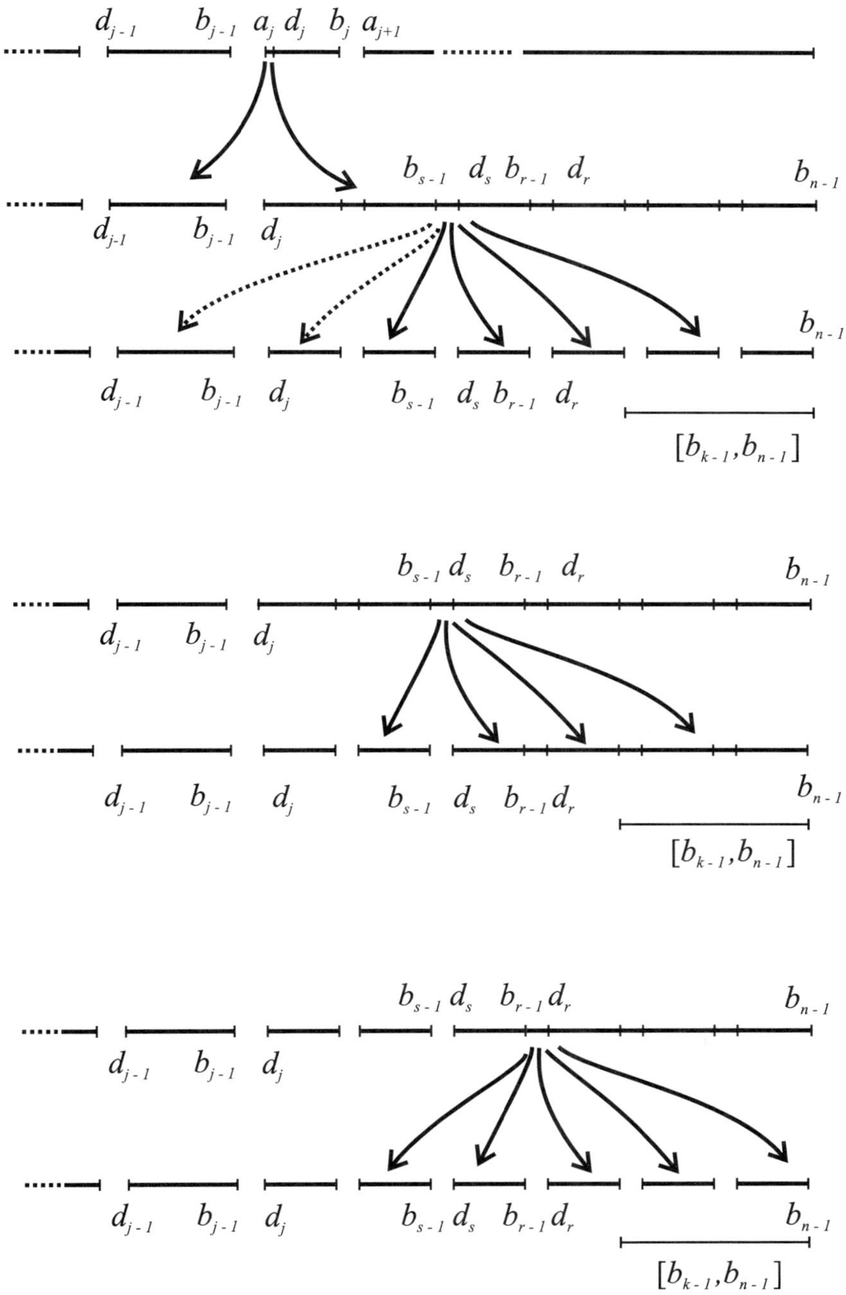

d_{j-1} b_{j-1} a_j d_j b_j a_{j+1}
b_{s-1} d_s b_{r-1} d_r b_{n-1}
d_{j-1} b_{j-1} d_j
b_{n-1}
d_{j-1} b_{j-1} d_j b_{s-1} d_s b_{r-1} d_r
[b_{k-1}, b_{n-1}]
b_{s-1} d_s b_{r-1} d_r b_{n-1}
d_{j-1} b_{j-1} d_j
d_{j-1} b_{j-1} d_j b_{s-1} d_s b_{r-1} d_r b_{n-1}
[b_{k-1}, b_{n-1}]
b_{s-1} d_s b_{r-1} d_r b_{n-1}
d_{j-1} b_{j-1} d_j
d_{j-1} b_{j-1} d_j b_{s-1} d_s b_{r-1} d_r b_{n-1}
[b_{k-1}, b_{n-1}]

Figure 21:

$$\leq \rho_j([b_{k-1}, b_{n-1}]) + \sum_{s=j+1}^{k-1} \mathrm{Bal}\Big(\rho_j\big|_{[b_{s-1},\, d_s]}, \cup_{l=s-1}^{k-1} I_l^* \cup [b_{k-1}, b_{n-1}]\Big)([b_{k-1}, b_{n-1}])$$

Let now

$$\rho_{j,s} = \mathrm{Bal}\Big(\rho_j\big|_{[b_{s-1},\, d_s]}, [d_{s-1}, b_{s-1}] \cup [d_s, b_{n-1}]\Big).$$

The the s-th term in the last sum is at most

$$\rho_{j,s}([b_{k-1}, b_{n-1}]) + \sum_{r=s+1}^{k-1} \rho_{j,s}([b_{r-1}, d_r]).$$

Summing up we obtain

$$\begin{aligned}
&\frac{M_{n,0}}{\sqrt{2^n}T}\Big(f_k(0,\ldots,0,\overbrace{1}^{k+1},0,0,\ldots,0) - f_k(1,\ldots,1,1,\overbrace{1}^{k+1},0,0,\ldots,0)\Big) \\
&\leq \sum_{j=1}^{k-1} \rho_j([b_{k-1}, b_{n-1}]) + \sum_{j=1}^{k-1}\sum_{s=j+1}^{k-1} \rho_{j,s}([b_{k-1}, b_{n-1}]) \\
&+ \sum_{j=1}^{k-1}\sum_{s=j+1}^{k-1}\sum_{r=s+1}^{k-1} \rho_{j,s}([b_{r-1}, d_r]) \\
&= S_1 + S_2 + S_3.
\end{aligned}$$

Using (6.58) and (6.59) we can write for $j \leq k-1$

$$\mu_{(K_n \setminus [a_k, d_k))}([a_j, d_j]) \leq \mu_{K^*_{n,k}}([a_j, b_j]) \leq C_0 \mu_{K_n}([a_j, b_j]),$$

where in the last step we used (6.53) (with k replaced by j). Thus for the density of ρ_j we get from (6.84) (with k replaced by j) for $t \in [3 \cdot 3^{j-1}, 3 \cdot 2^{n-1}]$

$$\frac{d\rho_j(t)}{dt} \leq C_0 \frac{32eM_{n,j}}{\sqrt{2^n}} \sqrt{\theta_j}\sqrt{\tau_j 2^j}\frac{64\sqrt{2}}{\pi}\frac{\sqrt{\theta_j\tau_j}}{t^{3/2}}\sqrt{2^j} < 3000C_0 \frac{M_{n,0}}{\sqrt{2^n}}\frac{\theta_j}{t^{3/2}}\sqrt{2^j} \tag{6.87}$$

by the choice of τ_j in (6.50). Therefore,

$$S_1 \leq 3000C_0 \frac{M_{n,0}}{\sqrt{2^n}} \sum_{j=1}^{k-1} \frac{2\theta_j}{\sqrt{2^{k-1}}}\sqrt{2^j} < 2 \cdot 10^4 C_0 \frac{M_{n,0}}{\sqrt{2^n}}\theta.$$

Next (6.87) and (6.85) give for the density of $\rho_{j,s}$, $s > j$, for $t \in [3 \cdot 3^{s-1}, 3 \cdot 2^{n-1}]$

$$\begin{aligned}
\frac{d\rho_{j,s}(t)}{dt} &\leq 3000C_0 \frac{M_{n,0}}{\sqrt{2^n}} \frac{\theta_j(2\theta_s 2^s)}{\sqrt{(2^{s-1})^{3/2}}}\sqrt{2^j}\frac{128}{\pi}\frac{\theta_s}{t^{3/2}}\sqrt{2^s} \\
&\leq 10^6 C_0 \frac{M_{n,0}}{\sqrt{2^n}}\frac{\theta_j\theta_s^2}{t^{3/2}}\sqrt{2^j}.
\end{aligned} \tag{6.88}$$

Therefore in view of $\theta_j < 1/16$

$$\begin{aligned}
S_2 &\leq \sum_{j=1}^{k-1}\sum_{s=j+1}^{k-1} 10^6 C_0 \frac{M_{n,0}}{\sqrt{2^n}}\theta_j\theta_s^2 \frac{2\sqrt{2^j}}{\sqrt{2^k}} \\
&= \sum_{s=2}^{k-1}\sum_{j=1}^{s-1} 10^6 C_0 \frac{M_{n,0}}{\sqrt{2^n}}\theta_j\theta_s^2\frac{2\sqrt{2^j}}{\sqrt{2^k}} \leq 2 \cdot 10^7 C_0 \frac{M_{n,0}}{\sqrt{2^n}}\theta^3 < 800C_0\theta.
\end{aligned}$$

Finally, for S_3 we get from (6.88)

$$\begin{aligned} S_3 &\leq \sum_{j=1}^{k-1}\sum_{s=j+1}^{k-1}\sum_{r=s+1}^{k-1} 10^6 C_0 \frac{M_{n,0}}{\sqrt{2^n}}\theta_j\theta_s^2\sqrt{2^j}\frac{2\theta_r 2^r}{(2^{r-1})^{3/2}} \\ &\leq 10^7 C_0 \frac{M_{n,0}}{\sqrt{2^n}}\sum_{p=1}^{\infty}\sum_{q=1}^{\infty}\sum_{j=1}^{k-1}\theta_j\theta_{j+p}^2\theta_{j+p+q}\frac{\sqrt{2^j}}{\sqrt{2^{j+p+q}}}, \end{aligned}$$

with the understanding that if $j+p\geq k$ then set $\theta_{j+p}=0$ and if $j+p+q\geq k$ then set $\theta_{j+p+q}=0$ in the last sum. By Hölder's inequality and (6.24)

$$\sum_{j=1}^{k-1}\theta_j\theta_{j+p}^2\theta_{j+p+q}\leq\theta\sum_{l}\theta_l^3\leq\theta^2,$$

therefore

$$S_3\leq 10^7 C_0\frac{M_{n,0}}{\sqrt{2^n}}\theta^2\sum_{p=1}^{\infty}\sum_{q=1}^{\infty}\frac{1}{\sqrt{2^p}}\frac{1}{\sqrt{2^q}}<10^6 C_0\frac{M_{n,0}}{\sqrt{2^n}}\theta.$$

All these show that

$$\begin{aligned} f_k(0,\dots,0,\overbrace{1}^{k+1},0,0,\dots,0) &- f_k(1,\dots,1,1,\overbrace{1}^{k+1},0,0,\dots,0) \\ &< 2\cdot 10^6 C_0 T\theta < 2\cdot 10^{16} C_0\theta, \end{aligned}$$

where C_0 is the constant from (6.59).

This completes the proof of (6.74) and with it the proof of Proposition 6.8.

■

Completion of the proof of Theorem 6.2

Exactly as in the proof of Theorem 6.5 we may assume that $\Theta(t)=t$ for $t\in[0,2]$.

Let for each $n=1,2,\dots$ the vector $\underline{\mathbf{x}}=\underline{\mathbf{x}}_n=(x_{n,0},\dots,x_{n,n-1})$ be the one guaranteed in Corollary 6.7, and let $E_n=E_{\underline{\mathbf{x}}_n}$ be the corresponding set. Then

$$E_n=\bigcup_{j=0}^{n-1}[2^j+x_{n,j}\tau_j 2^j, 2^{j+1}-\theta_{j+1}2^{j+1}].$$

If $g_{\mathbf{C}\setminus E_n}$ denotes the Green function of the complement of E_n and $T=T_n$ is the number from (6.69), then we set

$$h_n(z)=\exp\left(\frac{\sqrt{2^n}}{M_{n,0}}T_n g_{\mathbf{C}\setminus E_n}+iv_n(t)\right),$$

where $v_n(t)$ is the harmonic conjugate of

$$\frac{\sqrt{2^n}}{M_{n,0}}T_n g_{\mathbf{C}\setminus E_n}.$$

We may choose v_n to satisfy $v_n(0) = 0$. According to the discussion at the beginning of this section, the property that for any subinterval J of E_n the value

$$\frac{\sqrt{2^n}}{M_{n,0}} T_n \mu_{E_n}(J)$$

is an integer implies that $h_n(z)$ is a single valued analytic function on $\mathbf{C} \setminus E_n$. Let $J_{n,0}, \ldots, J_{n,m_n}$ be the subintervals of E_n numbered from the left to right. Note that if $k_1 < k_2 < \ldots$ are those indices k for which $\theta_k > 0$, then for $j = 1, \ldots, m_n - 1$

$$J_{n,j} = [a_{k_j} + x_{n,k_j} \tau_{k_j} 2^{k_j}, b_{k_{j+1}}],$$

i.e. the right endpoint of $J_{n,j}$ is fixed for all n, and the left endpoint lies in the interval $[a_{k_j}, a_{k_j} + \tau_{k_j} 2^{k_j}]$. Furthermore $J_{n,0} = [1, b_{k_1 - 1}]$ is also fixed for all n. All these imply that by repeatedly selecting subsequences we can find in a standard way an infinite subsequence $\mathcal{N}$ of the natural numbers for which the limits

$$\lim_{n \to \infty,\ n \in \mathcal{N}} x_{n,k_j} = x_{k_j}$$

exist for all j.

With $k_0 = \tau_0 = 0$, $a_0 = 1$ let

$$E = \bigcup_{j=0}^{\infty} [a_{k_j} + x_{k_j} \tau_{k_j} 2^{k_j}, b_{k_{j+1}}].$$

First of all we mention that this set satisfies $\Theta_E(t) \le \Theta(t)$ for all $t \ge 2$. In fact, by (6.23) and $\tau_j \le \theta_j \le \theta_j^*$ we get

$$\begin{aligned} |[2^j, 2^{j+1}] \setminus I_j(x_j)| &\le \tau_j 2^j + \theta_{j+1} 2^{j+1} \le \theta_j 2^j + \theta_{j+1} 2^{j+1} \\ &\le \frac{1}{8} \left[\Theta(2^{j-2}) - \Theta(2^{j-3}) + \Theta(2^{j-1}) - \Theta(2^{j-2}) \right] \\ &= \frac{1}{8} \left[\Theta(2^{j-1}) - \Theta(2^{j-3}) \right], \end{aligned}$$

and so for $t \in [2^k, 2^{k+1}]$

$$\begin{aligned} \Theta_E(t) &\le \sum_{j \le k} \frac{1}{8} \left[\Theta(2^{j-1}) - \Theta(2^{j-3}) \right] \le \frac{1}{8} \left[\Theta(2^{k-1}) + \Theta(2^{k-2}) \right] \\ &\le \frac{1}{4} \Theta(2^{k-1}) \le \frac{1}{4} \Theta(t). \end{aligned}$$

It follows from (6.66) and (6.70) that

$$\exp\left(\frac{10^9}{100}\right) \le h_n(0) \le \exp\left(2000 C_0 10^{10}\right). \tag{6.89}$$

Thus, $\log |h_n(z)|$ are positive harmonic functions on the sets $\mathbf{C} \setminus E_n$, and the preceding inequality and Harnack's inequality tells us that $h_n(z)$ are uniformly bounded analytic functions on every compact subset of $\mathbf{C} \setminus E$. We can therefore invoke the Arzela-Ascoli theorem in the effect that a subsequence of $\{h_n(z)\}_{n \in \mathcal{N}}$ converges on every compact subset of $\mathbf{C} \setminus E$. We may assume without loss of generality that

$h_n(z) \to f(z)$ locally uniformly in $\mathbf{C} \setminus E$ as $n \to \infty$, $n \in \mathcal{N}$. Then $f(z)$ is a single valued analytic function, and we show that it has the properties set forth in Theorem 6.2.

We claim first of all (6.2). Let J be a closed interval lying in the interior of one of the subintervals of E. Then for large $n \in \mathcal{N}$ this J belongs to E_n, hence on J we have $|h_n(z)| = 1$. Let $C_R(0)$ be a circle lying in $\mathbf{C} \setminus E$ and containing J in its interior. For any $z \in C_R(0)$ we have the estimate

$$\log|h_n(z)| \le \left(\max_{|w|=R} \log|h_n(w)|\right) \omega(z, C_R(0), \Delta_R(0) \setminus J),$$

hence the uniform boundedness of $\log|h_n|$ on C_R implies that if $\varepsilon > 0$ is given, then we can select a neighborhood G of J such that for $z \in G$ we have $\log|h_n(z)| \le \varepsilon$ for all large $n \in \mathcal{N}$. This gives $|f(z)| \le e^{\varepsilon}$ for all $z \in G$, and since J was an arbitrary subinterval, (6.2) follows.

The inequality (6.89) yields $|f(0)| \ge \exp(10^7)$, so $|f(0)| > 1$ is also true.

Finally we verify (6.3). Exactly as in the proof of Theorem 6.5, each $\log|h_n|$ takes its largest value at $-r$ on every circle $C_r(0)$ ($r < n$), which property is then inherited to $\log|f|$, i.e. $\log|f(z)| \le \log|f(-|z|)|$. Again Harnack's inequality shows

$$\log|f(-re^{it})| \ge \frac{1}{6} \log|f(-r)| \tag{6.90}$$

for $t \in [-\pi/2, \pi/2]$. Now let G be a neighborhood of $E \cap \Delta_r(0)$ with smooth boundary ∂G. Since

$$\begin{aligned}
\log|f(0)| &= \int_{C_r(0) \cup \partial G} (\log|f|) d\mathrm{Bal}\Big(\delta_0, C_r \cup \partial G\Big) \\
&\ge \int_{C_r(0)} (\log|f|) d\mathrm{Bal}\Big(\delta_0, C_r \cup \partial G\Big) \\
&\ge \frac{1}{6} \log|f(-r)| \int_{C_r^-(0)} \mathrm{Bal}\Big(\delta_0, C_r \cup \partial G,)\Big) \\
&= \frac{1}{6} (\log|f(-r)|) \omega(0, C_r^-(0), \Delta_r(0) \setminus G),
\end{aligned}$$

because $\log|f(t)|$ is, as being the limit of nonnegative functions, nonnegative. Here $C_r^-(0)$ denotes the left half of the circle $C_r(0)$. Now if we make G shrink to E then we can deduce

$$\log|f(0)| \ge \frac{1}{6} (\log|f(-r)|) \omega(0, C_r^-, \Delta_r(0) \setminus E),$$

and here, by (6.15), the last term on the right hand side is at least as large as $\omega(0, C_r^-(0), \Delta_r(0) \setminus E)/2$. All in all

$$\frac{\log|f(z)|}{|z|^{1/2}} \le \frac{\log|f(-r)|}{r^{1/2}} \le \frac{12 \log|f(0)|}{r^{1/2} \omega(0, C_r(0), \Delta_r(0) \setminus E)}.$$

Now we have already seen it in (6.13) that here the left hand side tends to 0 as $r \to \infty$. In fact, that proof in (6.13) used only the fact that

$$\int^{\infty} \frac{\Theta_E(t)}{t^3} dt = \infty,$$

which is true in the present case because $\sum_j \theta_j^2 = \infty$ (see Lemma 11.3 where one has to make the change of variables $t \to u/R$).

This verifies (6.3), and the proof of Theorem 6.2 is complete.

■

7 Markov and Bernstein type inequalities

In this chapter we give an application of our results to polynomial inequalities. Markov's inequality

$$\|P_n'\|_{[0,1]} \le 2n^2\|P_n\|_{[0,1]}, \qquad \deg(P_n) \le n, \tag{7.1}$$

is one of the basic inequalities connecting on an interval the size of the derivative of a polynomial P_n of degree at most n with its supremum norm $\|P_n\|_{[0,1]}$. If E is a compact set on the plane the analogue of Markov's inequality, namely

$$\|P_n'\|_E \le Cn^2\|P_n\|_E, \qquad \deg(P_n) \le n, \tag{7.2}$$

may or may not hold depending on the structure of E. For Cantor type sets we have a complete answer:

Theorem 7.1 *Let $\{\varepsilon_j\}$ be a sequence of numbers from the interval $[0,1)$. The Markov inequality*

$$\|P_n'\|_{\mathcal{C}(\varepsilon_1,\varepsilon_2,\dots)} \le Cn^2\|P_n\|_{\mathcal{C}(\varepsilon_1,\varepsilon_2,\dots)}, \qquad \deg(P_n) \le n, \tag{7.3}$$

is true with some constant C if and only if $\sum \varepsilon_j^2 < \infty$.

In particular, the Markov inequality (7.2) may hold even for sets of measure zero.

For general compact sets our results give a local version of Markov's inequality. Naturally, if the conditions of the next theorem hold uniformly for the points of a set E, then one gets a global Markov inequality on that set.

Theorem 7.2 *Let E be a compact set on the plane, $S \in E$, and let $\Theta_{E,S}(t)$ be the function* (2.1) *with respect to E and the point S. If*

$$\int_0^1 \frac{\Theta_{E,S}^2(u)}{u^3}\,du < \infty, \tag{7.4}$$

is true, then at S the local Markov inequality

$$|P_n'(S)| \le Cn^2\|P_n\|_E, \qquad \deg(P_n) \le n, \tag{7.5}$$

is true with some constant C.

Conversely, if Θ is a nonnegative increasing function on $[0,1]$ with the properties $\Theta(t) \le t$ for all t and

$$\int_0^1 \frac{\Theta^2(u)}{u^3}\,du = \infty, \tag{7.6}$$

then there is a compact set $E \subseteq [0,1]$ such that

$$\Theta_{E,0}(t) \le \Theta(t) \qquad \text{for all} \quad t \in [0,1], \tag{7.7}$$

and for some polynomials P_n of degree $n = 1,2,\dots$ we have

$$|P_n'(0)| \ne O\Big(n^2\|P_n\|_E\Big). \tag{7.8}$$

This whole work has emerged from Markov-type inequalities. T. Erdélyi, A. Kroó and J. Szabados [18] proved several local versions of the Markov inequality. They have also introduced the density function $\Theta_{E,S}$ and with it they proved an inequality which implies (7.5) provided

$$\int_0^1 \frac{\sqrt{\Theta_{E,S}(u)}}{u^2} du < \infty, \tag{7.9}$$

is true. However, this result is not sharp (note that $\Theta_{E,S}(u) \leq u$, hence the integrand in (7.9) is always bigger than the one in (7.4)), and the original motivation for this work was to find the correct form of the integrand. As an example consider for $0 < \alpha$ and $\alpha + 1 < \beta$ the set

$$E = \bigcup_{k=2}^{\infty} \left[\frac{1}{k^\alpha} + \frac{1}{k^\beta}, \frac{1}{(k-1)^\alpha}\right].$$

In this case we have

$$\Theta_{E,0}(t) \sim \sum_{k^{-\alpha} \leq t} \frac{1}{k^\beta} \sim t^{(\beta-1)/\alpha},$$

hence the aforementioned result in [18] is applicable exactly when $\beta > 2\alpha + 1$, and so it implies that for $\beta > 2\alpha + 1$ the local Markov inequality

$$|P_n'(0)| \leq Cn^2 \|P_n\|_E \tag{7.10}$$

is true. However, Theorem 7.2 gives that (7.10) is actually true for all $\beta > \alpha + 1$, and it is even true for the thinner set

$$E = \bigcup_{k=2}^{\infty} \left[\frac{1}{k^\alpha} + \frac{1}{k^{\alpha+1}\log k}, \frac{1}{(k-1)^\alpha}\right].$$

Indeed, for this set

$$\Theta_{E,0}(t) \sim \sum_{k^{-\alpha} \leq t} \frac{1}{k^{\alpha+1}\log k} \sim \frac{t}{\log 1/t},$$

hence the integral in (7.4) is finite for $S = 0$.

The following symmetric variant of Theorem 7.2 tells us under what conditions the analogue of the Bernstein inequality is true for general compact sets.

Theorem 7.3 *Let $E \subset [-1,1]$ be a compact set, and let $\Theta_E^*(t) = |[-t,t] \setminus E|$ be the function* (4.1). *If*

$$\int_0^1 \frac{(\Theta_E^*(u))^2}{u^3} du < \infty, \tag{7.11}$$

is true, then the Bernstein inequality

$$|P_n'(0)| \leq Cn\|P_n\|_E \qquad \deg(P_n) \leq n, \tag{7.12}$$

is true with some constant C.

Conversely, if Θ^* is a nonnegative increasing function on $[0,1]$ with the properties $\Theta^*(t) \leq 2t$ for all t and

$$\int_0^1 \frac{\Theta^*(u)^2}{u^3} du = \infty, \tag{7.13}$$

then there is a compact set $E \subseteq [-1,1]$ *such that*

$$\Theta_E^*(t) \le \Theta^*(t) \qquad \text{for all } \ t \in [0,1], \tag{7.14}$$

and for some polynomials P_n *of degree* $n = 1, 2, \dots$ *we have*

$$|P_n'(0)| \ne O\Big(n\|P_n\|_E\Big). \tag{7.15}$$

The local results are not convenient to use in the above form, so we state a form that is more general than the one given above. In fact, the same proof that gives (7.5) gives that if (7.4) is true, then for all $R > 0$ and $m = 1, 2, \dots$ we have

$$|P_n^{(m)}(z)| \le C_R m! n^{2m} \|P_n\|_{E \setminus \Delta_{Rn^{-2}}(S)}, \quad z \in \Delta_{Rn^{-2}}(S), \qquad \deg(P_n) \le n, \tag{7.16}$$

with some constant C_R, where

$$\Delta_{Rn^{-2}}(S) = \{z \,|\, |z - S| \le Rn^{-2}\}$$

is the disk about S of radius Rn^{-2}. In a similar manner, if (7.11) is true, then for $R > 0$ and $m = 1, 2, \dots$

$$|P_n^{(m)}(z)| \le C_R m! n^m \|P_n\|_{E \setminus [-R/n, R/n]}, \qquad |z| \le R/n, \quad \deg(P_n) \le n, \tag{7.17}$$

with a constant C_R. Finally, we shall also need the following strengthening of (7.3) about the origin: If $\sum_i \varepsilon_i^2 < \infty$, then for $R > 0$ and $m = 1, 2, \dots$

$$|P_n^{(m)}(z)| \le C_R m! n^{2m} \|P_n\|_{C(\varepsilon_1, \varepsilon_2, \dots) \setminus [0, Rn^{-2}]}, \quad |z| \le Rn^{-2}, \qquad \deg(P_n) \le n, \tag{7.18}$$

with some C_R depending on R.

7.1 Proof of Theorems 7.1, 7.2 and 7.3

First we prove that (7.4) implies (7.5). It follows from Theorem 2.2 that (7.4) implies

$$g_{\mathbf{C} \setminus E}(z) \le C|z - S|^{1/2}.$$

Thus, in view of the Bernstein-Walsh lemma (2.20)

$$|P_n(z)| \le e^{n g_{\mathbf{C} \setminus E}(z)} \|P_n\|_E,$$

we can see that for a polynomial P_n of degree at most n we have

$$|P_n(z)| \le e^{Cn|z - S|^{1/2}} \|P_n\|_E.$$

On applying this for $|z - S| = 1/n^2$, it follows that on the circle $|\xi - S| = 1/n^2$ we have

$$|P_n(\xi)| \le e^C \|P_n\|_E,$$

and thus Cauchy's formula for the derivative yields

$$|P_n'(S)| = \left| \frac{1}{2\pi i} \int_{|\xi - S| = 1/n^2} \frac{P_n(\xi)}{(\xi - S)^2} d\xi \right| \le e^C n^2 \|P_n\|_E.$$

The proof that in case of $\sum_j \varepsilon_j^2 < \infty$ the Markov type inequality (7.3) is true is identical, if we use the fact that by Theorem 5.1, in this case the Green function of $\mathbf{C} \setminus \mathcal{C}(\varepsilon_1, \varepsilon_2, \ldots)$ satisfies a (global) Lip 1/2 condition on $\mathcal{C}(\varepsilon_1, \varepsilon_2, \ldots)$.

Also, the proof of (7.12) is along similar lines, just use Corollary 4.1 instead of Theorem 2.2.

The stronger forms (7.16) and (7.17) follow from similar arguments. In fact, by Theorem 2.2 condition (7.4) implies

$$g_{\mathbf{C}\setminus(E\setminus\Delta_{2Rn^{-2}}(S))}(z) \le \frac{C_R}{n}$$

for all $|z - S| \le 2Rn^{-2}$. Now apply the Bernstein-Walsh lemma 2.20 and Cauchy's formula

$$|P_n^{(m)}(z)| = m! \left| \frac{1}{2\pi i} \int_{|\xi - z| = Rn^{-2}} \frac{P_n(\xi)}{(\xi - S)^{m+1}} d\xi \right|$$

exactly as before to deduce (7.16). The inequality (7.17) can be obtained in the same manner.

With (7.18) one has to be somewhat more careful, for we need an estimate for the Green function

$$g_{\mathbf{C}\setminus(\mathcal{C}(\varepsilon_1,\varepsilon_2,\ldots)\setminus[0,Rn^{-2}]},$$

and such estimates do not follow from Theorem 5.1. However, the appropriate estimate has been proven in Corollary 5.10. In fact, consider the numbers t_n from (5.3), select the largest s for which $t_s \ge 2Rn^{-2}$, and apply Corollary 5.10 to deduce that

$$g_{\mathbf{C}\setminus(\mathcal{C}(\varepsilon_1,\varepsilon_2,\ldots)\setminus[0,2Rn^{-2}]} \le \frac{C_R}{n},$$

and from here the proof is the same as that of (7.16).

Next we prove the second part of Theorem 7.2. Let Θ be as in the theorem, and consider the set E constructed in the proof of Theorem 3.1, i.e.

$$E = \bigcup_{k=1}^{\infty} \left[\frac{1}{2^k} + \frac{1}{4}\left(\Theta\left(\frac{1}{2^k}\right) - \Theta\left(\frac{1}{2^{k+1}}\right)\right), \frac{2}{2^k} \right]. \tag{7.19}$$

Let μ_k be the equilibrium measure of the set $E \cup [0, 2^{-k}]$. Then μ_k is absolutely continuous on $(0, 2^{-k})$ with respect to Lebesgue measure, and let S_k be the largest constant for which

$$\frac{d\mu_k(t)}{dt} \ge \frac{S_k}{\sqrt{t}} \quad \text{for } t \in \left(0, \frac{3}{4}2^{-k}\right] \tag{7.20}$$

(c.f. (3.11)). It was proved in the proof of Theorem 3.1 that if (7.6) holds, then

$$S_k \to \infty \qquad \text{as } k \to \infty \tag{7.21}$$

(see (3.14)).

It is clearly enough to show that if R_k is the largest constant for which there are polynomials $P_n \not\equiv 0$ of arbitrary large degree n with

$$|P_n'(0)| \ge R_k n^2 \|P_n\|_{E\cup[0,2^{-k}]}, \tag{7.22}$$

then $R_k \to \infty$. We shall do that by bounding R_k from below by an expression of S_k.

The set $E \cup [0, 2^{-k}]$ consists of a finite number of intervals, say

$$E = \cup_{j=1}^{l} [a_{2j-1}, a_{2j}], \qquad a_j < a_{j+1},\ a_1 = 0.$$

Its equilibrium measure has the form (see (2.17))

$$\frac{d\mu_E(t)}{dt} = \frac{\prod_{j=1}^{l-1} |t - \tau_j|}{\pi \prod_1^{2l} |(t - a_j)|^{1/2}} \tag{7.23}$$

for $t \in E$, where the numbers τ_j lie in the intervals (a_{2j}, a_{2j+1}). Now it was verified in [38, Theorem 4.1, (4.7)] that there is a sequence $\{P_n\}$ of polynomials of corresponding degree at most $n = 1, 2, \ldots$ such that

$$|P_n'(0)| \geq (1 + o(1)) 2 \frac{\prod_{i=1}^{l-1} (0 - \tau_i)^2}{\prod_{i \neq 1} |0 - a_i|} n^2 \|P_n\|_E,$$

where the factor $o(1)$ tends to zero as n tends to infinity. This shows that the R_k in (7.22) satisfies

$$R_k \geq \frac{\prod_{i=1}^{l-1} \tau_i^2}{\prod_{i \neq 1} a_i}. \tag{7.24}$$

Notice that if we write the equilibrium measure in (7.23) in the form

$$\frac{d\mu_E(t)}{dt} = \frac{Q_k(t)}{\sqrt{t}},$$

then the factor on the right hand side of (7.24) is nothing else than $(\pi Q_k(0))^2$, where $Q_k(0) := \lim_{t \to 0} Q_k(t)$. Thus,

$$R_k \geq \pi^2 Q_k(0)^2,$$

and since the definition of S_k in (7.20) implies that $Q_k(t) \geq S_k$ for $t \in \left(0, \frac{3}{4} 2^{-k}\right]$, we get $Q_k(0) \geq S_k$, and together with it the inequality

$$R_k \geq \pi^2 S_k^2.$$

Now $R_k \to \infty$ as $k \to \infty$ follows from (7.21), and this completes the proof of the second half of Theorem 7.2.

The necessity in Theorem 7.1 follows in a similar manner. In fact, let us deal only with the case $\sup \varepsilon_j < 1$, for the case $\sup \varepsilon_j = 1$ is even easier. Consider the set

$$\tilde{C}(\varepsilon_1, \varepsilon_2, \ldots) = [0, 1] \setminus \left(\bigcup_{k=1}^{\infty} (t_k, t_k + \varepsilon_k t_{k-1}) \right)$$

from (5.2). It is clearly enough to show that if $\sum_j \varepsilon_j^2 = \infty$, then there is a sequence $\{P_n\}$ of polynomials of corresponding degree $n = 1, 2, \ldots$ such that

$$|P_n'(0)| \neq O\Big(n^2 \|P_n\|_{\tilde{C}(\varepsilon_1, \varepsilon_2, \ldots)}\Big).$$

This $\tilde{C}(\varepsilon_1, \varepsilon_2, \ldots)$ has the same structure as the set (7.19), and we can repeat the preceding proof with little modification. In fact, the only modification we have to make is to use the sequence $\{t_n\}$ from (5.3) instead of the sequence $\{2^{-n}\}$. Indeed, in the proof of Corollary 3.2, it was proved that for $t \in (0, (t_{k+1} + t_k)/2]$ the inequality

$$\frac{d\mu_k(t)}{dt} \geq \frac{S_k}{\sqrt{t}},$$

holds, where now μ_k is the equilibrium measure of the set $\tilde{C}(\varepsilon_1, \varepsilon_2, \ldots) \cup [0, t_k]$, and for S_k we have by (3.16) the estimate

$$S_k \geq c \exp\left(d \int_{t_k}^1 \frac{\Theta^2(u)}{u^3} du \right) \to \infty \qquad \text{as } k \to \infty.$$

The rest is identical with the preceding proof.

Finally, we verify the second part of Theorem 7.3. Suppose that (7.13) holds, and consider the set appearing in Corollary 4.2. This corollary shows that if $a > 0$ is positive, then the density $\omega_{E\cup[-a,a]}(0)$ of the equilibrium measure $\mu_{E\cap[-a,a]}$ of the set $E \cup [-a, a]$ tends to infinity at the origin as $a \to 0$. Now (7.15) follows from the fact that according to Theorems 3.2 and 3.3 of [38] we have

$$\pi \omega_{E\cup[-a,a]}(0) = \sup_{P_n} \frac{|P_n'(0)|}{n\|P_n\|_{E\cup[-a,a]}}.$$

■

8 Fast decreasing polynomials

In this chapter we give an application of the main results to fast decreasing or so called pin polynomials. These are "Dirac delta" like polynomials, i.e. polynomials that decrease fast away from the origin. They play a significant role in several disciplines of mathematical analysis such as approximation theory, orthogonal polynomials, moment problems, etc. These applications require polynomials P that take the value 1 at the origin and are fast decreasing on $[-1,1]\setminus\{0\}$ in the sense that

$$P(0)=1, \quad |P(x)|\le e^{-\varphi(x)}, \quad x\in[-1,1], \tag{8.1}$$

where φ is a given even function that typically involves the degree of P. For example, the Markov-Bernstein-type inequality

$$\|wR'\|_{L^p(-\infty,\infty)}\le C\|wR\|_{L^p(-\infty,\infty)}, \qquad R \text{ a polynomial},$$

with an absolute constant C for weights like $w(x)=\exp(-|x|^\beta)$, $0<\beta<1$, follows from the existence of polynomials P_n of degree at most n satisfying

$$P_n(0)=1, \quad |P_n(x)|\le C\exp(-(n|x|)^\beta), \quad x\in[-1,1]$$

(see [26]).

The importance of such polynomials lies in the fact that their integrals provide good approximation to the signum function and thereby they serve as the building blocks for well localized "polynomial partition of unity". They are also optimal convolution kernels, for they imitate the Dirac delta as close as possible.

The problem can be formulated in two different ways: find the fastest decreasing polynomials of a given order, or alternatively, find the smallest possible degree for the polynomial P in (8.1). This degree is denoted by n_φ.

The order of n_φ can be estimated by an explicitly computable quantity as follows (see [20]): Let φ be an even function, right continuous and increasing on $[0,1]$. Then

$$\frac{1}{6}N_\varphi\le n_\varphi\le 12N_\varphi$$

where $N_\varphi=0$ if $\varphi(1)\le 0$ and

$$N_\varphi=2\sup_{\varphi^{-1}(0)\le x<b}\sqrt{\frac{\varphi(x)}{x^2}}+\int_b^{1/2}\frac{\varphi(x)}{x^2}dx+\sup_{1/2\le x<1}\frac{\varphi(x)}{-\log(1-x)}+1,$$

$b=\min(\varphi^{-1}(1),1/2)$, otherwise.

Here, for $u\ge 0$,

$$\varphi^{-1}(u)=\sup\{\tau\,|\,\tau\in[0,1],\ \varphi(\tau)\le u\}.$$

If $N_\varphi=\infty$, then the statement of the theorem means that there are no polynomials whatsoever with the stated properties.

As a special case the following holds. Let φ be an even and on $[0,1]$ increasing function with $\varphi(0)=\varphi(0+0)=0$ and $\varphi(x)\le C\varphi(x/2)$ for $x\in[0,1]$. Then there are polynomials P_n of degree at most n satisfying

$$P_n(0)=1, \quad |P_n(x)|\le D\exp(-dn\varphi(x)), \quad x\in[-1,1], \quad n=0,1,\ldots \tag{8.2}$$

for some constants $D > 0$ and $d > 0$ if and only if

$$\int_0^1 \frac{\varphi(u)}{u^2} du < \infty. \tag{8.3}$$

For example,

$$|P(0)| = 1, \quad |P(x)| \leq C_1 \exp(-n|x|^\beta)$$

with $\beta > 1$ can be achieved by polynomials of degree $\leq Cn$, but for $\varphi(x) = |x|$ we get that the minimal degree n_φ of the polynomials P satisfying

$$|P(0)| = 1, \quad |P(x)| \leq C_1 e^{-n|x|}, \quad x \in [-1, 1]$$

satisfies

$$\frac{1}{C} n \log n \leq n_\varphi \leq Cn \log n.$$

The substitution $x \to x^2$ changes the problem into fast decreasing polynomials on $[0, 1]$. E.g. in this case (8.2) takes the form

$$P_n(0) = 1, \qquad |P_n(x)| \leq De^{-dn\varphi(x)} \quad x \in [0, 1], \tag{8.4}$$

and for this (8.3) will change into the necessary and sufficient condition

$$\int_0^1 \frac{\varphi(t)}{t^{3/2}} dt < \infty. \tag{8.5}$$

Now suppose, that $E \subset [0, 1]$ is a closed set, and we want

$$|P_n(x)| \leq De^{-dn\varphi(x)}$$

only on E (besides the property $P_n(0) = 1$), i.e. we care only for the decrease of P_n on E. It is clear that a thinner E will allow faster decrease for P_n, and we are going to show that the denseness condition used in this work is precisely the condition under which a full analogue of (8.4)–(8.5) holds.

Theorem 8.1 *Let $E \subseteq [0, 1]$ be compact, and φ an increasing function on $[0, 1]$ with $\varphi(0) = 0$. If*

$$\int_0^1 \frac{\Theta_E(t)^2}{t^3} dt < \infty, \tag{8.6}$$

then for $n = 1, 2, \ldots$ there are polynomials with

$$P_n(0) = 1, \qquad |P_n(x)| \leq C_1 e^{-c_1 n\varphi(x)}, \quad x \in E \tag{8.7}$$

with some constants $C_1, c_1 > 0$ if and only if

$$\int_0^1 \frac{\varphi(t)}{t^{3/2}} dt < \infty. \tag{8.8}$$

Conversely, if $0 \leq \Theta(t) \leq t$ is an increasing function on $[0, 1]$ with

$$\int_0^1 \frac{\Theta(t)^2}{t^3} dt = \infty, \tag{8.9}$$

then there are a compact set $E \subset [0, 1]$ such that $\Theta_E(t) \leq \Theta(t)$ for all t and a monotone φ with

$$\int_0^1 \frac{\varphi(t)}{t^{3/2}} dt = \infty \tag{8.10}$$

for which there are polynomials P_n with property (8.7).

By the standard symmetrization $x \to x^2$ one can immediately deduce the following corollary.

Corollary 8.2 *Let $E \subseteq [-1,1]$ be compact, and φ an even function on $[-1,1]$ such that φ is increasing on $[0,1]$ and $\varphi(0)=0$. Let furthermore*

$$\Theta_E^*(t) = |[-t,t] \setminus E|$$

be the symmetric density function from (4.1). *If*

$$\int_0^1 \frac{\Theta_E^*(t)^2}{t^3} dt < \infty, \tag{8.11}$$

then for $n = 1, 2, \ldots$ there are polynomials with

$$P_n(0) = 1, \qquad |P_n(x)| \le C_1 e^{-c_1 n \varphi(x)}, \quad x \in E \tag{8.12}$$

with some constants $C_1, c_1 > 0$ if and only if

$$\int_0^1 \frac{\varphi(t)}{t^2} dt < \infty. \tag{8.13}$$

Conversely, if $0 \le \Theta^*(t) \le 2t$ is an increasing function on $[0,1]$ with

$$\int_0^1 \frac{\Theta^*(t)^2}{t^3} dt = \infty, \tag{8.14}$$

then there are a compact set $E \subset [-1,1]$ such that $\Theta_E^*(t) \le \Theta^*(t)$ for all t and a monotone φ with

$$\int_0^1 \frac{\varphi(t)}{t^2} dt = \infty \tag{8.15}$$

for which there are polynomials P_n with property (8.12).

For Cantor type sets we prove the following. Let $\varepsilon_i \in [0,1)$, and consider the Cantor set $\mathcal{C}(\varepsilon_1, \varepsilon_2, \ldots)$ formed with these numbers.

Theorem 8.3 *Let φ be an increasing function on $[0,1]$ with $\varphi(0) = 0$ and $E = \mathcal{C}(\varepsilon_1, \varepsilon_2, \ldots)$. If $\sum_i \varepsilon_i^2 < \infty$, then there are polynomials with the property* (8.7) *if and only if* (8.8) *is true.*

Conversely, if $\sum_i \varepsilon_i^2 = \infty$, then there is a φ with (8.10) *for which there are polynomials P_n with the property* (8.7) *for $E = \mathcal{C}(\varepsilon_1, \varepsilon_2, \ldots)$.*

Proof of the first part of Theorem 8.1. Let us suppose that (8.6) is true, and that there are polynomials P_n with property (8.7). We have to prove (8.8). To this end first we estimate the size of P_n on $[0,1] \setminus E$, and since this will be done by the Bernstein-Walsh lemma (2.20), first we estimate the Green function $g_{\mathbf{C} \setminus E}$ there. Thus, let $x \in [2^{-k}, 2^{-k+1}]$, and let $C_{2^{-k-1}}(x)$ be the circle about x of radius 2^{-k-1}. If $y \in C_{2^{-k-1}}(x)$, then we get from Theorem 2.2 that

$$g_{\mathbf{C} \setminus (E \cap [2^{-k-1}, 1])}(y) \le C 2^{-k/2}. \tag{8.16}$$

On the other hand, from Lemma 2.5 we obtain

$$\omega(x, C_{2^{-k-1}}(x), \Delta_{2^{-k-1}}(x) \setminus E) \le C 2^k \Theta_E(2^{-k+2}). \tag{8.17}$$

A combination of these two give

$$\begin{aligned} g_{\mathbf{C}\setminus(E\cap[2^{-k-1},1])}(x) &\le \omega(x, C_{2^{-k-1}}(x), \Delta_{2^{-k-1}} \setminus (E\cap[2^{-k-1},1])) \times \\ &\quad \times \max_{y\in C_{2^{-k-1}}(x)} g_{\mathbf{C}\setminus(E\cap[2^{-k-1},1])}(y) \\ &\le C_2 2^{k/2}\Theta_E(2^{-k+2}). \end{aligned}$$

Since on the set $E\cap[2^{-k-1},1]$ we have

$$|P_n(x)| \le C_1 \exp\Big(-c_1 n\varphi(2^{-k-1})\Big),$$

it follows from the Bernstein-Walsh lemma (2.20) that for $x\in[2^{-k},2^{-k+1}]\setminus E$ we have

$$|P_n(x)| \le C_1 \exp\Big(-c_1 n\varphi(2^{-k-1}) + C_2 n 2^{k/2}\Theta_E(2^{-k+2})\Big). \tag{8.18}$$

Since $P_n(0)=1$, we can write P_n in the form

$$P_n(x) = \prod_{j=1}^{n}(1-\alpha_j x),$$

and by writing instead of α_j its absolute value if necessary, and then writing instead of any $\alpha_j<1$ the value $\alpha_j=1$ (these operations decrease the modulus of P_n on $[0,1]$), we may assume without loss of generality that all $\alpha_j\ge 1$. Now we use (see [36, (3.5)]) that for $\alpha\ge 1$

$$\int_0^1 \frac{\log|1-\alpha t|}{t^{3/2}\sqrt{1-t}}\,dt = -2\pi,$$

which implies

$$\int_a^1 \frac{\log|1-\alpha t|}{t^{3/2}\sqrt{1-t}}\,dt \ge -2\pi$$

for all $\alpha\ge 1$ and $a>0$. Thus, we obtain from (8.7) and (8.18)

$$\begin{aligned} -2n\pi &\le \int_a^1 \frac{\log|P_n(t)|}{t^{3/2}\sqrt{1-t}}\,dt = \int_{E\cap[a,1]} + \int_{[a,1]\setminus E} \\ &\le \int_{E\cap[a,1]} \frac{\log C_1 - c_1 n\varphi(t)}{t^{3/2}\sqrt{1-t}}\,dt \\ &\quad + \sum_{k=1}^{[\log_2 1/a]+1} \Big(\log C_1 - c_1 n\varphi(2^{-k-1}) + C_2 n 2^{k/2}\Theta_E(2^{-k+2})\Big) \times \\ &\qquad \times \int_{[2^{-k},2^{-k+1}]\setminus E} \frac{1}{t^{3/2}\sqrt{1-t}}\,dt \end{aligned}$$

For the very last integral we have the bound

$$\int_{[2^{-k},2^{-k+1}]\setminus E} \frac{1}{t^{3/2}\sqrt{1-t}}\,dt \le 2^{3k/2}\Theta_E(2^{-k+1}),$$

therefore we get

$$-2n\pi \le \int_a^1 \frac{\log C_1}{t^{3/2}\sqrt{1-t}}dt + \int_a^1 \frac{-c_1 n\varphi(t/4)}{t^{3/2}\sqrt{1-t}}dt + 32C_2 n \sum_k \Theta_E(2^{-k})^2 2^{2k}.$$

Here the last sum is at most a constant times the integral in (8.6), hence it is bounded, say less than C_3. Now on letting $n \to \infty$ we obtain

$$c_1 \int_a^1 \frac{\varphi(t/4)}{t^{3/2}\sqrt{1-t}}dt \le 2\pi + 32C_2C_3,$$

and on letting $a \to \infty$ we obtain (8.8).

That conversely, if (8.8) is true than there are polynomials with the property (8.7) follows from the fact that, as we have already mentioned, the same is true for $E = [0,1]$. This completes the proof of the first part of Theorem 8.1.

■

Proof of the second part of Theorem 8.1. Let us suppose that (8.9) holds, and consider the set E from (3.10). For $\tau > 0$ set

$$E_\tau = (E \cup [0, \gamma_\tau]) \setminus [0, \tau), \tag{8.19}$$

where $0 < \gamma_\tau < 1$ is a number to be chosen below in such a way that $\gamma_\tau \to 0$ as $\tau \to 0$. We get from Theorem 3.1, (3.5) and from Harnack's inequality that

$$\begin{aligned} g_{\mathbf{C}\setminus E_\tau}(0) &\ge \frac{1}{3} g_{\mathbf{C}\setminus E_\tau}(-\tau) \ge \frac{1}{3} g_{(\mathbf{C}\setminus E)\cup[0,\gamma_\tau]}(-\tau) \\ &\ge c\sqrt{\tau}\exp\left(d\int_{\gamma_\tau}^1 \frac{\Theta_{E\cup[0,\gamma_\tau]}(u)^2}{u^3}du\right) = 2h(\tau)\sqrt{\tau}, \end{aligned} \tag{8.20}$$

where $h(\tau)$ is a function that depends on γ_τ and that tends to infinity as $\tau \to 0$, because $\gamma_\tau \to 0$. We may also assume that $h(\tau)\sqrt{\tau}$ is an increasing function of τ. For fixed τ the set E_τ consists of a finite number of intervals and we have the representation

$$g_{\mathbf{C}\setminus E_\tau}(z) = \log \frac{1}{\operatorname{cap}(E_\tau)} - U^{\mu_{E_\tau}}(z),$$

where the equilibrium measure μ_{E_τ} is given by a density function that is positive and continuous inside E_τ and has a $\sim 1/\sqrt{|t-a|}$ behavior at any endpoint a of a subinterval of E_τ. Now there is a well known technique to discretize such a logarithmic potential, see e.g. Theorem [30, Theorem VI.4.1]. However, this is a subtle and important point in the proof, hence we included into the Appendix the relevant result. Thus, according to Theorem 11.1 from the Appendix for every $\nu \ge 0$, ν not necessarily an integer, there are polynomials $Q_\nu^* = Q_{\nu,\tau}^*$ of degree at most ν with all their zeros lying in E_τ such that with some constant $M(\gamma_\tau)$ independent of ν

$$Q_\nu^*(0) \ge e^{\nu g_{\mathbf{C}\setminus E_\tau}(0)} \ge e^{2\nu h(\tau)\sqrt{\tau}}, \tag{8.21}$$

and

$$|Q_\nu^*(t)| \le M(\gamma_\tau) \qquad \text{for} \quad t \in E_\tau. \tag{8.22}$$

Since all the zeros of Q^*_ν lie in E_τ, we also have

$$0 \le Q^*_\nu(t) \le Q^*_\tau(0) \qquad \text{for} \quad 0 \le t \le \tau.$$

Now if the number $\gamma_\tau \to 0$ goes to zero very slowly, then $M(\gamma_\tau)$ tends to infinity as slowly as we wish (consider e.g. piecewise constant γ_τ), hence we may choose $\gamma_\tau \to 0$ in such a way that $M(\gamma_\tau) \le M_1/\tau$ for all $\tau > 0$ with some fixed constant $M_1 \ge 1$. Now for the polynomials

$$|Q_\nu(t)| = Q_{\nu,\tau}(t) = \frac{Q^*_{\nu,\tau}(t)}{Q^*_{\nu,\tau}(0)},$$

we have

$$Q_{\nu,\tau}(0) = 1, \qquad 0 \le Q_{\nu,\tau}(t) \le 1 \quad \text{for} \quad 0 \le t \le \tau$$

and for $t \in E$, $t \ge \tau$

$$Q_{\nu,\tau}(t) \le \frac{M_1}{\tau} e^{-3\nu h(\tau)\sqrt{\tau}},$$

which gives in particular

$$Q_{\nu,\tau}(t) \le e^{-\nu h(\tau)\sqrt{\tau}} \qquad \text{if} \quad \nu h(\tau)\sqrt{\tau} \ge \log \frac{M_1}{\tau}. \tag{8.23}$$

Since this estimate is true only for $\nu h(\tau)\sqrt{\tau} \ge \log M_1/\tau$, we also need the modified Chebyshev polynomials, for which a similar, but slightly weaker inequality is true, but for all ν and τ with $\nu\sqrt{\tau} \ge 1$. Consider the Chebyshev polynomials

$$T_\nu(x) = \cos(\nu \arccos x) = \frac{1}{2}\left\{\left(x + \sqrt{x^2-1}\right)^\nu + \left(x - \sqrt{x^2-1}\right)^\nu\right\},$$

for which $|T_\nu(x)| \le 1$ is true for all $x \in [-1, 1]$. Let

$$T_{\nu,\tau}(t) = T_{[\nu]+1}\left(\frac{1+\tau-2t}{1-\tau}\right) \Big/ T_{[\nu]+1}\left(\frac{1+\tau}{1-\tau}\right).$$

For these it easily follows that $T_{\nu,\tau}(0) = 1$, $|T_{\nu,\tau}(t)| \le 1$ for $0 \le t \le 1$, and for $\tau \le t \le 1$

$$\begin{aligned} |T_{\nu,\tau}(t)| &\le T_{[\nu]+1}\left(\frac{1+\tau}{1-\tau}\right)^{-1} = 2\left(\left(\frac{1+\sqrt{\tau}}{1-\sqrt{\tau}}\right)^{[\nu]+1} + \left(\frac{1-\sqrt{\tau}}{1+\sqrt{\tau}}\right)^{[\nu]+1}\right)^{-1} \\ &\le 2e^{-([\nu]+1)2\sqrt{\tau}} \le 2e^{-2\nu\sqrt{\tau}} \le e^{-\nu\sqrt{\tau}} \end{aligned} \tag{8.24}$$

provided $\nu\sqrt{\tau} \ge 1$.

After these preliminaries we can make the construction. Recall that we want to construct polynomials $P_0(t)$ of degree at most Cn that are $\le Ce^{-n\varphi(t)}$ on E, with an increasing continuous function $\varphi \ge 0$ for which (8.10) holds. We shall choose φ later, but in any case we set $\varphi(t) = t^2$ for $1/2 \le t \le 2$, and we shall have $\varphi^{-1}(t) \ge t^2$ for all $t \in [0, 1]$, where φ^{-1} denotes the inverse function. For $1 \le 2^k \le n$, k integer, choose

$$\tau_{k,n} = \varphi^{-1}\left(\frac{2^k}{n}\right),$$

and let

$$\nu_{k,n} = \frac{2^{k+1}}{\sqrt{\tau_{k,n}}h(\tau_{k,n})}$$

if $2^{k+1} \geq 3\log n$, and otherwise let

$$\nu_{k,n} = \frac{2^{k+1}}{\sqrt{\tau_{k,n}}}.$$

Note that in the first case we have for large n

$$\nu_{k,n}h(\tau_{k,n})\sqrt{\tau_{k,n}} = 2^{k+1} \geq 3\log n \geq \log\frac{M_1}{(1/n)^2} \geq \log\frac{M_1}{\varphi^{-1}(2^k/n)} = \log\frac{M_1}{\tau_{k,n}},$$

so that in this case for the polynomial $Q_{\nu_{k,n},\tau_{k,n}}(t)$ the inequality (8.23) holds. For other k we use the polynomial $T_{\nu_{k,n},\tau_{k,n}}(t)$, for which (8.24) is true because

$$\nu_{k,n}\sqrt{\tau_{k,n}} = 2^{k+1} > 1.$$

We write with these polynomials

$$P_0(t) = \prod_{2\leq 2^{k+1}\leq 3\log n} T_{\nu_{k,n},\tau_{k,n}}(t) \prod_{3\log n<2^{k+1}\leq 4n} Q_{\nu_{k,n},\tau_{k,n}}(t).$$

In what follows, besides the estimates (8.23) and (8.24), we shall also use that each factor is at most 1 in absolute value on the set E.

If $t \in E$, $\tau_{k,n} \leq t < \tau_{k+1,n}$, and here k is such that $2^{k+1} > 3\log n$, then we get from (8.23)

$$|P_0(t)| \leq |Q_{\nu_{k,n},\tau_{k,n}}(t)| \leq e^{-\nu_{k,n}h(\tau_{k,n})\sqrt{\tau_{k,n}}} = e^{-2^{k+1}} = e^{-n\varphi(\tau_{k+1,n})} \leq e^{-n\varphi(t)},$$

and similarly in case of $2^{k+1} \leq 3\log n$, we get from (8.24)

$$|P_0(t)| \leq T_{\nu_{k,n},\tau_{k,n}}(t) \leq e^{-\nu_{k,n}\sqrt{\tau_{k,n}}} = e^{-2^{k+1}} = e^{-n\varphi(\tau_{k+1,n})} \leq e^{-n\varphi(t)}.$$

If $t \in E$ and $t < \tau_{0,n} = \varphi^{-1}(1/n)$ (i.e. when t does not belong to any of the intervals $[\tau_{k,n}, \tau_{k+1,n}]$), then

$$|P_0(t)| \leq 1 = e\cdot e^{-n\varphi(\varphi^{-1}(1/n))} \leq e\cdot e^{-n\varphi(t)}.$$

These show that (8.7) is true for all $t \in E$.

For the degree of P_0 we have

$$\deg(P_0) = \sum_{2\leq 2^{k+1}\leq 3\log n} \frac{2^{k+1}}{\sqrt{\varphi^{-1}(2^k/n)}} + \sum_{3\log n<2^{k+1}\leq 4n} \frac{2^{k+1}}{\sqrt{\varphi^{-1}(2^k/n)h(\varphi^{-1}(2^k/n))}}.$$

If we multiply the sums by n and at the same time divide every term by n, then we get some Riemannian lower sums for the monotonic functions

$$\frac{1}{\sqrt{\varphi^{-1}(u)}} \quad \text{and} \quad \frac{1}{\sqrt{\varphi^{-1}(u)h(\varphi^{-1}(u))}},$$

and this way we can see that

$$\deg(P_0) \leq 4n \int_{1/2n}^{3\log n/n} \frac{du}{\sqrt{\varphi^{-1}(u)}} + 4n \int_{3\log n/2n}^{2} \frac{du}{\sqrt{\varphi^{-1}(u) h(\varphi^{-1}(u))}}. \tag{8.25}$$

Since $h(v) \to \infty$ as $v \to 0$, we can select a monotone φ^{-1} such that $\varphi^{-1}(t) \geq t^2$,

$$\int_v^{\sqrt{v}} \frac{du}{\sqrt{\varphi^{-1}(u)}} \to 0$$

as $v \to 0$,

$$\int_0^1 \frac{du}{\sqrt{\varphi^{-1}(u) h(\varphi^{-1}(u))}} < \infty,$$

but

$$\int_0^1 \frac{du}{\sqrt{\varphi^{-1}(u)}} = \infty.$$

Now substitute $u = \varphi(v)$, integrate by parts and use that $\varphi^{-1}(t) \geq t^2$ to conclude from the last relation

$$\int_0^1 \frac{\varphi(v)}{v^{3/2}} dv = \infty.$$

Thus, with such a choice we have $\deg(P_0) \leq Cn$ with some constant C (see (8.25)), and the polynomials P_0 have all the properties claimed in the second part of Theorem 8.1 (if we want P_0 to have degree at most n, then just replace n by $[n/C]$).

■

Proof of Theorem 8.3. Suppose first that $\sum_i \varepsilon_i^2 < \infty$. We have to show that if there are polynomials P_n with property (8.7), then (8.8) is true. We follow the corresponding part of the proof of Theorem 8.1. That proof was based on the fact that

$$\sum_{k=1}^{\infty} \int_{2^{-k}}^{2^{-k+1}} \frac{g_{\mathbf{C} \setminus (E \cap [2^{-k-1},1])}(u)}{u^{3/2}} du < \infty, \tag{8.26}$$

so it is enough to prove this for $E = \mathcal{C}(\varepsilon_1, \varepsilon_2, \ldots)$.

We use the notations and results from Chapter 5. Let t_k be the numbers introduced in (5.3). These will play the role of 2^{-k} in the proof of Theorem 8.1. Thus, instead of (8.26) it will be more convenient to verify

$$\sum_{k=1}^{\infty} \int_{t_k}^{t_{k-1}} \frac{g_{\mathbf{C} \setminus (E \cap [t_{k+1},1])}(u)}{u^{3/2}} du < \infty, \tag{8.27}$$

the rest of the proof is the same as above in the proof of Theorem 8.1.

For $x \in [t_k, t_k + 1] \setminus E$, set $r_x = x - t_{k+1}$, and let $C_{r_x}(x)$ be the circle about x and of radius r_x. The point x belongs to one of the contiguous intervals to

$\mathcal{C}(\varepsilon_1, \varepsilon_2, \ldots)$, say to a contiguous intervals omitted at the $k+m$-th level, i.e. $x \in \mathcal{C}_{k+m}(\varepsilon_1, \varepsilon_2, \ldots) \setminus \mathcal{C}_{k+m+1}(\varepsilon_1, \varepsilon_2, \ldots)$. The analogue of (8.16) is

$$g_{\mathbf{C}\setminus(E\cap[t_{k+1},1])}(y) \leq C\sqrt{t_k}, \qquad y \in C_{r_x}(x),$$

which follows from Corollary 5.10 and from $|y| \geq t_{k+1}$. By Lemma 5.8 and Theorem 5.9 (see (5.24)) the analogue of (8.17) in the present case is

$$\omega(x, C_{r_x}(x), \Delta_{r_x}(x) \setminus E) \leq C\sigma^m(\varepsilon_{k+m+1} + \sum_{j>k+m+1} \varepsilon_j^2 \sigma^{j-(k+m+1)})$$

with some $\sigma < 1$. Combining these two we obtain

$$g_{\mathbf{C}\setminus(E\cap[t_{k+1},1])}(x) \leq C\sqrt{t_k}\sigma^m(\varepsilon_{k+m+1} + \sum_{j>k+m+1} \varepsilon_j^2 \sigma^{j-(k+m+1)}).$$

If I denotes the contiguous interval containing x, then $|I| = \varepsilon_{k+m+1}t_{k+m}$, and hence the just obtained estimate gives

$$\int_I \frac{g_{\mathbf{C}\setminus(E\cap[t_{k+1},1])}(x)}{x^{3/2}}dx \leq C\sqrt{t_k}\sigma^m(\varepsilon_{k+m+1} + \sum_{j>k+m+1} \varepsilon_j^2 \sigma^{j-(k+m+1)}) \times$$
$$\times\, t_k^{-3/2}\varepsilon_{k+m+1}t_{k+m}.$$

There is one such I if $m = 0$, and there are 2^{m-1} such I's if $m \geq 1$ (in other words, $[t_{k+1}, t_k] \cap \Big(\mathcal{C}_{k+m}(\varepsilon_1, \varepsilon_2, \ldots) \setminus \mathcal{C}_{k+m+1}(\varepsilon_1, \varepsilon_2, \ldots)\Big)$ consists of that many intervals), thus

$$\int_{t_k}^{t_{k-1}} \frac{g_{\mathbf{C}\setminus(E\cap[t_{k+1},1])}(x)}{x^{3/2}}dx \leq \sum_{m=0}^{\infty} 2^m C\sqrt{t_k}\sigma^m \times$$
$$\times\, (\varepsilon_{k+m+1} + \sum_{j>k+m+1} \varepsilon_j^2\sigma^{j-(k+m+1)})t_k^{-3/2}\varepsilon_{k+m+1}t_{k+m},$$

which, in view of $t_{k+m} \leq t_k/2^m$ yields

$$\sum_{k=1}^{\infty}\int_{t_k}^{t_{k-1}} \frac{g_{C\setminus(E\cap[t_{k+1},1]}(u)}{u^{3/2}}du \leq C\sum_{k=0}^{\infty}\sum_{m=0}^{\infty}\sigma^m\varepsilon_{k+m+1} \times$$
$$\times\, (\varepsilon_{k+m+1} + \sum_{j>k+m+1} \varepsilon_j^2\sigma^{j-(k+m+1)}),$$

and it is easy to see (in fact, we have seen it several times in the proofs in Chapter 5) that the right hand side is finite here. This proves (8.27), and the first part of Theorem 8.3 has been verified.

To prove the second part of the theorem let us assume $\sum_i \varepsilon_i^2 = \infty$. First we show that the case $\sup_i \varepsilon_i = 1$ can be reduced to the case $\sup_i \varepsilon_i < 1$. To this end it is sufficient to show that $\mathcal{C}(\varepsilon_1, \varepsilon_2, \ldots)$ is contained in a $\mathcal{C}(\varepsilon_1^*, \varepsilon_2^*, \ldots)$ with all $\varepsilon_i^* \leq 7/9$ and $\sum_i(e_i^*)^2 = \infty$. In fact, let $S(\varepsilon_i)$ be the operation applied to a set of intervals Σ that omits the middle $\varepsilon_i|I|$ of every interval I in Σ. Then the Cantor construction is nothing else than repeated application of the operations $S(\varepsilon_1)$, $S(\varepsilon_2), \cdots$ (starting from $\Sigma = [0,1]$). Now if for some i we have $\varepsilon_i > 7/9$, say

$$\frac{1}{3^{m+1}} \leq \frac{1-\varepsilon_i}{2} < \frac{1}{3^m}$$

with some $m \geq 2$, then replace the operation $S(\varepsilon_i)$ by $(m-1)$ successive application of $S(1/3)$, and one application of $S(1-(1-\varepsilon_i)3^{m-1})$. On a subinterval I of $\mathcal{C}_{i-1}(\varepsilon_1, \varepsilon_2, \ldots)$ these latter operations produce 2^m intervals of length $|I|(1-\varepsilon_i)/2$, among which we find the two intervals of $S(\varepsilon_i)$. Since we have $1-(1-\varepsilon_i)3^{m-1} \leq 1-2/9 = 7/9$, we may replace $S(\varepsilon_i)$ by the aforementioned operations with parameters all lying in the interval $[0, 7/9]$, and in doing so we enlarge our Cantor set. It is also clear that if $\sup_i \varepsilon_i = 1$, then we have to do this replacement infinitely many times, so $\sum_i (\varepsilon_i^*)^2 = \infty$ is automatically satisfied.

Thus, without loss of generality we may assume $\varepsilon_i \leq \varepsilon < 1$ and $\sum_i \varepsilon_i^2 = \infty$. Consider the numbers t_k and from (5.3). For them we have $t_k(1-\varepsilon)/2 \leq t_{k+1} \leq t_k/2$, i.e. the condition (3.6) holds. Now let $E = [0,1] \setminus \left(\cup_{k=1}^{\infty}(t_k, t_k^*)\right)$, where $t_k^* = t_k(1+\varepsilon_k)/(1-\varepsilon_k)$ are the numbers from (5.4). In other words, E is the set that we obtain if during the Cantor construction at each step we omit the middle interval only from the leftmost remaining interval. It follows from Corollary 3.3 that

$$\lim_{r \to 0+} g_{\mathbf{C} \setminus E}(-r)/\sqrt{r} = \infty,$$

furthermore the proof of that corollary even gives for $\gamma_r \to 0$

$$\lim_{r \to 0+} g_{\mathbf{C} \setminus (E \cup [0, \gamma_r])}(-r)/\sqrt{r} = \infty, \tag{8.28}$$

no matter how slowly γ_r tends to 0 as $r \to 0$.

After these preliminaries we follow again the proof of Theorem 8.1, this time the second part. That proof was based on the fact that the set E_τ from (8.19) consists of finitely many intervals, and that $g_{\mathbf{C} \setminus E_\tau}(0)/\sqrt{\tau} \to \infty$ as $\tau \to 0$ (see (8.20)). (8.28) shows that these properties are also true in the present case, and we can copy the proof word for word to obtain a φ with (8.10) for which there are polynomials P_n with the property (8.7). Finally, all we have to mention is that $\mathcal{C}(\varepsilon_1, \varepsilon_2, \ldots) \subset E$.

■

9 Remez and Schur type inequalities

Remez-type inequalities estimate the norm of a polynomial. The simplest case of it says that if $|P_n(x)| \leq 1$ for $x \in [n^{-2}, 1]$, then with an absolute constant C we have $|P_n(x)| \leq C$ for $x \in [0, n^{-2}]$. Of course, this is just a simple appearance of the Bernstein-Walsh lemma (2.20) and of the $\sqrt{|z|}$ behavior of the Green function. Connected with the aforementioned inequality are the Schur type inequalities that estimate the norm of a polynomial provided its weighted norm is known. For example, if $\sigma > 0$, then

$$\|P_n\|_{[0,1]} \leq C n^{2\sigma} \|x^\sigma P_n(x)\|_{[0,1]}.$$

The results from the other chapters of this work give easy extensions of these summarized in

Theorem 9.1 *Let E be a compact subset of the plane and let us suppose that*

$$\int_0^1 \frac{\Theta_E^2(u)}{u^3} du < \infty, \tag{9.1}$$

holds. Then

$$|P_n(z)| \leq C_R \|P_n\|_{E \setminus \Delta_{Rn^{-2}}(0)}, \qquad |z| \leq Rn^{-2} \tag{9.2}$$

where C_R depends only on R; furthermore if $\sigma > 0$, then

$$\|P_n\|_E \leq C_\sigma n^{2\sigma} \|z^\sigma P_n(z)\|_E. \tag{9.3}$$

Conversely, if $0 \leq \Theta(t) \leq t$ is an increasing function on $[0,1]$ with the property

$$\int_0^1 \frac{\Theta_E^2(u)}{u^3} du = \infty, \tag{9.4}$$

then there is a compact set $E \subset [0,1]$ such that $\Theta_E(t) \leq \Theta(t)$ for all $t \in [0,1]$, and for each n there are polynomials P_n of degree at most n such that

$$\frac{|P_n(0)|}{\|P_n\|_{E \cap [n^{-2},1]}} \to \infty, \qquad n \to \infty, \tag{9.5}$$

and for any $\sigma > 0$

$$\frac{\|P_n\|_E}{n^{2\sigma} \|x^\sigma P_n(x)\|_E} \to \infty, \qquad n \to \infty. \tag{9.6}$$

Similar proofs give that if $E = \mathcal{C}(\varepsilon_1, \varepsilon_2, \ldots)$ is a Cantor set, then the Remez type inequality (9.2) holds if and only if $\sum \varepsilon_i^2 < \infty$, and this is also the necessary and sufficient condition for the validity of the Schur inequality (9.3).

In connection with Remez type inequalities let us mention the work [2], where they are connected with logarithmic capacity.

Proof. Since (9.1) implies for $E_n = E \cap \{|z| \geq Rn^{-2}\}$ that

$$g_{\mathbf{C} \setminus E_n}(z) \leq C_R/n$$

for $|z| \le Rn^{-2}$ (see Theorem 2.2), the inequality (9.2) is an immediate consequence of the Bernstein-Walsh lemma (2.20). On applying (9.2) with $R = 1$ we get (9.3) since for $|z| \ge n^{-2}$ we have $|P_n(z)| \le n^{2\sigma}|z^\sigma P_n(z)|$.

Now suppose (9.4) is true, and let E be the set defined in (3.10). With a fixed k put $E_k = E \cup [0, 2^{-k}]$, for which we proved in (7.22) that for large m (say $m \ge m_k$) there are polynomials Q_m of degree at most m such that

$$Q'_m(0) \ge R_k m^2 \|Q_m\|_{E\cup[0,2^{-k}]}, \tag{9.7}$$

where $R_k \to \infty$ as $k \to \infty$. Let now $Q_{m,k}$ be a (actually unique) polynomial such that $\|Q_{m,k}\|_{E_k} \le 1$ and for which $Q'_{m,k}(0)$ is maximal among all polynomials of degree at most m with this property. Thus, for $m \ge m_k$ we have $Q'_{m,k}(0) \ge R_k m^2$, and it is clear that $\|Q_{m,k}\|_{E_k} = 1$. We claim that $Q_{m,n}(0) = -1$. In fact, if

$$Q_{m,k}(x) = a\prod_{j=1}^{m}(x - \alpha_j), \qquad a = a_{k,m}, \quad \alpha_j = \alpha_j^{k,m}$$

and not every α_j is positive, then we can decrease the norm $\|Q_{m,k}\|_{E_k}$of $Q_{m,k}$ by replacing every α_j by its absolute value, so we must have $\alpha_j \ge 0$ for all j. Let α_1 be the smallest zero of $Q_{m,k}$. Then $Q_{m,k}(x)$ is negative and monotone increasing for $x < \alpha_1$. Therefore, if we had $Q_{m,k}(0) > -1$, we could move α_1 to the right a little into α_1^* by keeping this property, i.e. for $Q^*_{m,k}(x) = Q_{m,k}(x)(x-\alpha_1^*)/(x-\alpha_1)$ we have $0 \ge Q^*_{m,k}(0) > -1$. But in this change then the norm of the polynomial on E_k decreases, i.e. $\|Q^*_{m,k}\|_{E_k} < 1$, while the value

$$Q'_{m,k}(0) = a_{m,k}\sum_{j=1}^{m}\prod_{i\ne j}(-\alpha_i)$$

increases, i.e. $(Q^*_{m,k})'(0) > Q'_{m,k}(0)$, which is impossible by the choice of $Q_{m,k}$. Thus, $Q_{m,k}(0) = -1$. Fix $m = m_k$ with $Q'_{m,k}(0) \ge R_k m^2$.

Now with the classical Chebyshev polynomials $T_l(x) = \cos(l \arccos x)$ set for large n

$$P_n^*(x) = (-1)^{[n/m]+1}T_{[n/m]}(Q_{m,k}(x)).$$

This is of degree at most n, and we have $\|P_n\|_{E_k} = 1$, $P_n^*(0) = -1$, and since $T_l'(-1) = (-1)^{l+1}l^2$,

$$(P_n^*)'(0) = Q'_{m,k}(0)[n/m]^2 \ge R_k m^2[n/m]^2 \ge R_k n^2/2.$$

We claim that for large n the polynomial P_n has at least two zeros in the interval $[1/4n^2, 1/2n^2]$. If n is large, then this interval is mapped by $Q_{m,k}$ into an interval which contains

$$[-1 + (4/3)Q'_{m,k}(0)(1/4n^2), -1 + (3/4)Q'_{m,k}(0)(1/2n^2)]$$

of length $\ge Q'_{m,k}(0)/24n^2 \ge R_k/24(n/m)^2$, and $T_{[n/m]}$ has at least

$$(\sqrt{R_k/24(n/m)^2}/8)[n/m] \ge \sqrt{R_k}/50$$

zeros on that interval, so P_n^* has at least two zeros in $[1/4n^2, 1/2n^2]$ if $R_k > 10^4$. But then there is a point $x_n \in [1/4n^2, 1/2n^2]$ with $|P_n^*(x_n)| = 1$, because $T_{[n/m]}$ takes at least one of the values ± 1 in between any two of its zeros.

Now let $\tilde{P}_n$ be the polynomial that we obtain from P_n^* by replacing each of its zeros in the interval $[0, 1/n^2]$ by one-one zero at $1/n^2$. On the set $[1/n^2, 1] \cap E$ this operation decreases the norm, thus $\|\tilde{P}_n\|_{[1/n^2,1]\cap E} \le 1$, and at the same time it increases the value of the derivative at the origin, i.e. $\tilde{P}_n'(0) \ge R_k n^2/2$. It is also clear that $|\tilde{P}_n(x_n)| \ge |P_n^*(x_n)| = 1$, and since $\tilde{P}_n$ is increasing and negative for $x < 1/n^2$, we can infer that $\tilde{P}_n(x) < -1$ for $x < 1/4n^2$ (recall that $1/4n^2 < x_n$). Therefore, we have $|\tilde{P}_n(z)| \ge 1$ for all $\Re z < 1/4n^2$ because all the zeros of $\tilde{P}_n$ lie in $[1/n^2, 1)$.

Since the derivative of $\tilde{P}_n$ is monotone decreasing for $x < 1/n^2$ (recall that all the zeros of $\tilde{P}_n$ lie in $[1/n^2, 1]$ and $(\tilde{P}_n)'(0) \ge 1$), we can conclude from $\tilde{P}_n'(0) \ge R_k n^2/2$ first that $\tilde{P}_n'(t) \ge R_k n^2/2$ for $t \in [-1/n^2, 0]$, and then that

$$\tilde{P}_n(-1/n^2) = \tilde{P}_n(0) - \int_{-1/n^2}^{0} \tilde{P}_n'(t)dt < -R_k/2.$$

Now we use that the function $\log|\tilde{P}_n(z)|$ is harmonic and nonnegative in $\Re z < 1/4n^2$. Therefore, we get from Harnack's inequality (2.29) applied to the disk $\Delta_{5/4n^2}(-1/n^2)$ that

$$|\tilde{P}_n(0)| \ge |\tilde{P}_n(-1/n^2)|^{1/10} \ge (R_k/2)^{1/10}.$$

This is true for all large n, and since R_k can be arbitrary large provided k is sufficiently small, we can see that $\tilde{P}_n(0)/\|\tilde{P}_n\|_{E\cap[n^{-2},1]}$ tends to infinity as $n \to \infty$. This proves (9.5).

The preceding proof also gives that if (9.4) holds then there are polynomials Q_n such that $\|Q_n\|_E = Q_n(0) = 1$ and for some sequence $c_n \searrow 0$ ($c_0 = 1$)

$$\|Q_n\|_{E\cap[c_n n^{-2},1]} \le c_n. \tag{9.8}$$

Let now for an n the number m be selected so that $n/2 < 2^{m+1} \le n$, and consider

$$P_n(x) = \prod_{j=1}^{m} Q_{2^j}(x).$$

This has degree at most n, and if $x \in E$, $c_{2^{k+1}} 2^{-2(k+1)} < x \le c_{2^k} 2^{-2k}$ for some $k < m$, then

$$\begin{aligned} n^{2\sigma} x^\sigma |P_n(x)| &\le 4^{2\sigma} 2^{2m\sigma} \left(\frac{c_{2^k}}{2^{2k}}\right)^\sigma c_{2^{k+1}} c_{2^{k+2}} \cdots c_{2^m} \\ &= 4^{2\sigma} c_{2^k}^\sigma (2^{2\sigma} c_{2^{k+1}})(2^{2\sigma} c_{2^{k+2}}) \cdots (2^{2\sigma} c_{2^m}) \le c_{2^m}, \end{aligned}$$

for large m (recall that $c_{2^k} \to 0$ as $k \to \infty$), while for $x \in E$, $0 < x \le c_{2^m} 2^{-2m}$

$$n^{2\sigma} x^s |P_n(x)| \le 4^{2\sigma} 2^{2m\sigma} \left(\frac{c_{2^m}}{2^{2m}}\right)^\sigma \le 4^{2\sigma} c_{2^m}^\sigma.$$

Thus,

$$n^{2\sigma} \|x^s P_n(x)\|_E \le 4^{2\sigma} c_{2^m}^{\min(1,2\sigma)} \to 0,$$

and this is (9.6).

Note that for the polynomials P_n we have just constructed (9.5) also holds (see (9.8)), and this completes the proof.

∎

10 Approximation on compact sets

The approximation of the $|x|$ function by polynomials is a key to many problems in approximation theory. For example, both the Weierstrass theorem and the Jackson theory can be based on it, and the problem is also fundamental in spline theory. Let $\mathcal{E}_n(f, [-1, 1])$ denote the error of best approximation to f on $[-1, 1]$ by polynomials of degree at most n. S. N. Bernstein [11] proved in 1914, that the limit

$$\lim_{n\to\infty} n\mathcal{E}_n(|x|, [-1, 1]) = \sigma \tag{10.1}$$

exists, finite and positive. This is a rather difficult result (with a proof over 50 pages). For σ he showed $0.278 < \sigma < 0.286$, and based on that he conjectured that $\sigma = (2\sqrt{\pi})^{-1}$, but that was disproved in [40] by high precision calculations, which also gave σ up to 50 decimal places. The exact value of σ is still unknown. Bernstein returned to the same problem in the period 1938-46 in the papers [12], [13] and he established that for $p > 0$, p not an even integer the finite and nonzero limits

$$\lim_{n\to\infty} n^p\mathcal{E}_n(|x|^p, [-1, 1]) = \sigma_p \tag{10.2}$$

$$\lim_{n\to\infty} n^p\mathcal{E}_n(\mathrm{sign}(x)|x|^p, [-1, 1]) = \sigma_p^*$$

exist, furthermore that for $x_0 \in (-1, 1)$

$$\lim_{n\to\infty} n^p\mathcal{E}_n(|x - x_0|^p, [-1, 1]) = (1 - x_0^2)^{p/2}\sigma_p \tag{10.3}$$

holds true, where σ_p is the same constant as in (10.2).

In this section we consider the problem what happens for more general sets. If $E \subset \mathbf{R}$ is compact, then let

$$\mathcal{E}_n(f, E) = \inf_{\deg(P_n)} \|f - P_n\|_E$$

be the best approximation of f on E by polynomials of degree at most n in the supremum norm. Recall that

$$\Theta_E^*(t) = |[-t, t] \setminus E|$$

is the symmetric density function for E.

Theorem 10.1 *Let $E \subset \mathbf{R}$ be compact. If*

$$\int_0^1 \frac{\Theta_E^*(t, x_0)^2}{t^3} dt < \infty, \tag{10.4}$$

then for $p > 0$ not an even integer we have

$$\liminf_{n\to\infty} n^p\mathcal{E}_n(|x - x_0|^p, E) > 0. \tag{10.5}$$

Conversely, if $0 \le \Theta^*(t) \le t$ is an increasing function on $[0, 1]$ with

$$\int_0^1 \frac{\Theta^*(t)^2}{t^3} dt = \infty, \tag{10.6}$$

then there is a compact set $E \subset [-1, 1]$ such that $\Theta_E^*(t, 0) \le \Theta^*(t)$ for all t and

$$\lim_{n\to\infty} n^p\mathcal{E}_n(|x|^p, E) = 0. \tag{10.7}$$

Without loss of generality we may assume $x_0 = 0$. It will be easier to work with approximation on $[0,1]$ of $|x|^{p/2}$ rather than with approximation on $[-1,1]$ of $|x|^p$. This can be easily done by considering $E^* = E \cap (-E)$, for which the nonsymmetric density function Θ_{E^*} satisfies

$$\Theta_{E^*}(t) \le \Theta_E^*(t) \le 2\Theta_{E^*}(t).$$

Furthermore the substitution $x \to x^2$ shows that

$$\mathcal{E}_n(|x|^p, E) = \mathcal{E}_{[n/2]}(|x|^{p/2}, E^*),$$

hence Theorem 10.1 is an immediate consequence of

Theorem 10.2 *Let $E \subset [0,\infty)$ be a compact set. If*

$$\int_0^1 \frac{\Theta_E(t)^2}{t^3}\,dt < \infty, \tag{10.8}$$

then for $p > 0$ not an integer we have

$$\liminf_{n\to\infty} n^{2p}\mathcal{E}_n(x^p, E) > 0. \tag{10.9}$$

Conversely, if $0 \le \Theta(t) \le t$ is an increasing function on $[0,1]$ with

$$\int_0^1 \frac{\Theta(t)^2}{t^3}\,dt = \infty, \tag{10.10}$$

then there is a compact set $E \subset [0,1]$ such that $\Theta_E(t) \le \Theta(t)$ for all t and

$$\lim_{n\to\infty} n^{2p}\mathcal{E}_n(x^p, E) = 0. \tag{10.11}$$

For approximating on the Cantor sets we prove

Theorem 10.3 *Let $\varepsilon_i \in [0,1)$, and consider the Cantor-type set $\mathcal{C}(\varepsilon_1, \varepsilon_2, \ldots)$ constructed with the sequence $\{\varepsilon_i\}$ (see Chapter 5). Then for $p > 0$ not an integer*

$$\liminf_{n\to\infty} n^{2p}\mathcal{E}_n(x^p, \mathcal{C}(\varepsilon_1, \varepsilon_2, \ldots)) > 0 \tag{10.12}$$

if and only if $\sum_i \varepsilon_i^2 < \infty$.

In particular, with the choice $\varepsilon_i = 1/(i+1)$ we obtain a set $E = \mathcal{C}(\varepsilon_1, \varepsilon_2, \ldots)$ of zero linear measure for which (10.12) is true. By symmetrization (i.e. considering $E = \{x \in [-1,1] \mid x^2 \in \mathcal{C}(1/2, 1/3, \ldots)\}$ we can deduce

Corollary 10.4 *There is a compact set $E \subset [-1,1]$ of zero linear measure such that for $p > 0$ not an even integer*

$$\liminf_{n\to\infty} n^{p}\mathcal{E}_n(|x|^p, E) > 0. \tag{10.13}$$

Before discussing the condition (10.4) let us mention that there are numerous results on approximation on general sets, the interested reader should consult the monographs [4] and [32] or the paper [3].

The density condition (10.4) appeared in the work [41] by R. K. Vasiliev in a slightly different form. Vasiliev considered approximation of $|x-x_0|^p$ on compact subsets, and his approach is as follows. Let

$$E = [-1,1] \setminus \cup_{i=1}^{\infty}(\alpha_i, \beta_i),$$

and consider the sets

$$E_m = [-1,1] \setminus \cup_{i=1}^{m-1}(\alpha_i, \beta_i).$$

E_m consists of m intervals

$$E_m = \cup_{j=1}^{m}[a_j, b_j]$$

$a_1 < b_1 < a_2 < b_2 \cdots b_{m-1} < a_m < b_m$, and for it define

$$h_{E_m}(x) = \frac{\prod_{j=1}^{m-1}|x-\lambda_j|}{\sqrt{\prod_{j=1}^{m}|x-a_j||x-b_j|}},$$

where λ_j are chosen so that

$$\int_{b_k}^{a_{k+1}} \frac{\prod_{j=1}^{m-1}(t-\lambda_j)}{\sqrt{\prod_{j=1}^{m}|t-a_j||t-b_j|}}dt = 0 \tag{10.14}$$

for all $k = 1, \ldots, m-1$. Now set

$$h_E(x) = \lim_{m\to\infty} h_{E_m}(x) = \sup_m h_{E_m}(x),$$

where it can be shown that the limit exists (but it is not necessarily finite).

Now with these notations Vasiliev claims the following two results:

$$\lim_{n\to\infty} n^p \mathcal{E}_n(|x-x_0|^p, E) = h_E(x_0)^{-p}\sigma_p, \tag{10.15}$$

$$\lim_{n\to\infty} n^p \mathcal{E}_n(|x-x_0|^p, E) > 0 \Longleftrightarrow \int_0^1 \frac{\Theta_E^*(t,x_0)^2}{t^3}dt < \infty. \tag{10.16}$$

This second claim contradicts Corollary 10.4 (for a set E of zero measure we have $\Theta_E^*(t,x_0) \equiv 2t$), and the correct relevant result is Theorem 10.1. Vasiliev's paper [41] is 166 pages long, and the proof of (10.15) and (10.16) is about 160 pages, so it is difficult to say what went wrong in the proof. We do not know if the full (10.15) is correct, but we have a relatively simple proof that shows its validity provided x_0 lies in the interior of E.

To formulate the result let us recall (2.17), according to which the density of the equilibrium measure for a set

$$E = \cup_{j=1}^{m}[a_j, b_j]$$

$a_1 < b_1 < a_2 < b_2 \cdots b_{m-1} < a_m < b_m$ is given by

$$\omega_E(x) = \frac{\prod_{j=1}^{m-1}|x-\lambda_j|}{\pi\sqrt{\prod_{j=1}^{m}|x-a_j||x-b_j|}}, \tag{10.17}$$

where λ_j are chosen so that (10.14) is true for all $k = 1, \ldots, m-1$. Thus, Vasiliev's function is just $h_E(x) = \pi\omega_E(x)$ if E consists of a finite number of intervals, and also if E is arbitrary compact, but x is in its interior (see Lemma 10.8 below). Now (10.15) for $x_0 \in \mathrm{Int}(E)$ takes the following form.

Theorem 10.5 (R. K. Vasiliev) *Let $E \subseteq \mathbf{R}$ be compact and let x_0 be a point in the interior of E. Then*

$$\lim_{n\to\infty} n^p \mathcal{E}_n(|x-x_0|^p, E) = (\pi\omega_E(x_0))^{-p}\sigma_p, \tag{10.18}$$

where σ_p is the constant from Bernstein's theorem (10.2).

For example, if $F = [-1,1]$, then

$$\pi\omega_{[-1,1]}(x) = \frac{1}{\sqrt{1-x^2}},$$

and in this special case we recapture Bernstein's result (10.3).

Proof of Theorem 10.2. First consider the case when (10.8) is true. This part of the proof has common elements with converse results in approximation theory. We argue by contradiction, so let us suppose that for some subsequence $\mathcal{N}_0$ of the natural numbers we have

$$\lim_{n\to\infty,\ n\in\mathcal{N}_0} n^{2p}\mathcal{E}_n(x^p, E) = 0.$$

By considering the largest power of 2 not exceeding n instead of n then it easily follows that for some $\mathcal{N} \subset \mathbf{N}$

$$\lim_{n\to\infty,\ n\in\mathcal{N}} 2^{2pn}\mathcal{E}_{2^n}(x^p, E) = 0.$$

Select and fix a number $m > p+1$. Let $\varepsilon > 0$ be an arbitrary small number, and suppose that $n \in \mathcal{N}$ is so large that $\mathcal{E}_{2^n}(x^p, E) < \varepsilon 2^{-2pn}$. Let P_{2^n} be the polynomial of best approximation of degree 2^n to x^p on E. Thus

$$|x^p - P_{2^n}(x)| \le \varepsilon 2^{-2pn} \qquad x \in E. \tag{10.19}$$

We also know (see (10.2)) that there is a constant C_0 such that for all j (not necessarily lying in the sequence $\mathcal{N}$)

$$|x^p - P_{2^j}(x)| \le C_0 2^{-2pj} \qquad x \in E. \tag{10.20}$$

The condition (10.8) easily implies that the set E has right density 1 at the origin, hence we can select a number $x_n \in [2^{-2n}, 2^{-2n+1)}]$ such that the numbers $x_n, 2x_n, 3x_n, \ldots, (m+1)x_n$ all belong to E. Set

$$\Delta_{n,m}f = x_n^{-m}\sum_{i=0}^{m}(-1)^i\binom{m}{i}f((m+1-i)x_n).$$

The sum on the right is called the m-th forward difference of f (with step size x_n), and it is well known [17, (4.7.4) and (4.7.9), pp. 120-121] that if f is m times continuously differentiable on the interval $[x_n, (m+1)x_n]$, then

$$\Delta_{n,m}f = f^{(m)}(\xi_n)$$

for some $\xi_n \in [x_n, (m+1)x_n]$ (that may depend on f). Now apply this with $f(x) = x^p$ and $n+s$ instead of n to conclude that

$$|\Delta_{n+s,m}f| = |(p(p-1)\cdots(p-m+1)||\xi_{n+s}^{p-m}| \ge c_0 2^{2(n+s)(m-p)} \tag{10.21}$$

with

$$c_0 = |(p(p-1)\cdots(p-m+1)|(m+1)^{p-m}.$$

On the other hand, we obtain from (10.19)

$$|\Delta_{n+s,m}(f - P_{2^n})| \le x_{n+s}^{-m} 2^m \varepsilon 2^{-2pn} \le 2^m 2^{2ps} \varepsilon 2^{2(n+s)(m-p)}. \tag{10.22}$$

Finally, we have

$$|\Delta_{n+s,m} P_{2^n}| = |P_{2^n}^{(m)}(\xi)|$$

with some $\xi \in [x_{n+s}, (m+1)x_{n+s}]$. Here

$$|P_{2^n}^{(m)}(\xi)| \le \sum_{j=0}^{n-1} |(P_{2^{j+1}} - P_{2^j})^{(m)}(\xi)|,$$

and by applying (7.16) to the terms in the sum we obtain

$$|(P_{2^{j+1}} - P_{2^j})^{(m)}(\xi)| \le C_1 (2^{j+1})^{2m} \|P_{2^{j+1}} - P_{2^j}\|_E.$$

But by (10.20)

$$\|P_{2^{j+1}} - P_{2^j}\|_E \le \|P_{2^{j+1}} - f\|_E + \|P_{2^j} - f\|_E \le 2C_0 2^{-2pj},$$

hence we can conclude

$$\begin{aligned} |\Delta_{n+s,m} P_{2^n}| &\le 2C_0 C_1 2^{2m} \sum_{j=0}^{n-1} 2^{2(m-p)j} \le 4C_0 C_1 2^{2m} 2^{2n(m-p)} \\ &= 4C_0 C_1 2^{2m} 2^{-s(m-p)} 2^{2(n+s)(m-p)}. \end{aligned}$$

However, this contradicts (10.21) and (10.22) if we choose ε and s in such a way that

$$4C_0 C_1 2^{2m} 2^{-s(m-p)} + 2^m 2^{2ps} \varepsilon < c_0.$$

This contradiction proves the first part of Theorem 10.2.

To prove the second part of Theorem 10.2 let us suppose that (10.10) holds, and consider the set E from (3.10). We shall show that for this set (10.11) is true. To simplify the proof, let us consider the concrete value $p = 1/2$ of p; for other p's the idea is the same.

We shall frequently use Jackson's theorem (see e.g. [17, Theorem 6.6.2, p. 219]): if g is an m-times differentiable function on an interval J and its m-th derivative belongs to a Lip α class, then

$$\mathcal{E}_n(g, J) \le C n^{-m-\alpha},$$

i.e. g can be approximated in the order $n^{-m-\alpha}$ by polynomials of degree at most n.

Let $a > 0$ be a small number, and set $h_a(x) = \sqrt{x}$ if $x \ge a$ and $h_a(x) = \sqrt{a} - (x-a)/2\sqrt{a}$ if $0 \le x \le a$. The function h_a is continuously differentiable and its derivative is in the Lip 1 class (with a Lipshitz constant depending on a), therefore on any interval we can approximate it by polynomials of degree at most n in the order $O_a(1/n^2)$, and so the order of approximation is at most $n^{-3/2}$ if $n \ge n_a$.

Therefore it is enough to prove that the order of approximation of $\sqrt{x} - h_a(x)$ on E by polynomials of degree at most n is at most $c_a n^{-1}$, where c_a tends to zero as $a \to 0$.

The function

$$\begin{aligned} h^*(t) &= \sqrt{t} - h_a(t600a)/\sqrt{600a} \\ &= \begin{cases} \sqrt{t} - \frac{1}{\sqrt{600}} + \frac{1}{2\sqrt{600}}(600t - 1) & \text{if } 0 \le t \le 1/600 \\ 0 & \text{if } 1/600 \le t \le 1 \end{cases} \end{aligned}$$

can be approximated on $[0,1]$ by polynomials of degree at most m in the order $\le C_1/m$ (this follows e.g. from (10.1)). With some number N_a depending on a and tending to infinity as $a \to 0$ we set $m = \sqrt{a}N_a n$. Thus, there are polynomials $S_{\sqrt{a}N_a n}$ of degree at most $\sqrt{a}N_a n$ with

$$|h^*(t) - S_{\sqrt{a}N_a n}(t)| \le \frac{C_1}{\sqrt{a}N_a n}. \tag{10.23}$$

On setting here $t = u/600a$ and multiplying through by $\sqrt{600a}$ we obtain

$$\left|(\sqrt{u} - h_a(u)) - S_{\sqrt{a}N_a n}(u/600a)\sqrt{600a}\right| \le \frac{C_1\sqrt{600}}{N_a n} \tag{10.24}$$

for all $0 \le u \le 600a$. In particular, the polynomial $S_{\sqrt{a}N_a n}(u/600a)\sqrt{600a}$ is at most 2 in absolute value on the interval $[0, 600a]$, and so it follows from the Bernstein-Walsh lemma (2.24) that for all $t \in [600a, 1]$ we have

$$|S_{\sqrt{a}N_a n}(t/600a)\sqrt{600a}| \le 2\left(\frac{4t}{600a}\right)^{\sqrt{a}N_a n} \le 2\exp\left(nN_a\sqrt{a}\log\frac{t}{a}\right). \tag{10.25}$$

Thus, if we can construct polynomials $P_0 = P_{0,a,n}$ of degree at most n such that for all $t \in [0,1]$ we have $0 \le P_0(t) \le 1$,

$$|1 - P_0(t)| \le n^{-2} \qquad \text{for} \quad t \in [0, a] \tag{10.26}$$

and

$$P_0(t) \le \exp\left(-2n\sqrt{a}N_a \log\frac{t}{a}\right) \qquad \text{for} \quad t \in E,\ t \ge 600a, \tag{10.27}$$

then for the polynomial $P_0^* = P_0(t)S_{\sqrt{a}N_a n}(u/600a)\sqrt{600a}$ of degree at most $2n$ we have for $t \in [0,1]$ the estimate

$$|(\sqrt{t} - h_a(t)) - P_0^*(t)| \le \frac{C_1\sqrt{600}}{N_a n} + e^{-nN_a\sqrt{a}} + \frac{2}{n^2}. \tag{10.28}$$

In fact, for $t \in [0,a]$ this follows from (10.24) and (10.26), for $t \in [a, 600a]$ from (10.24) and from the facts that $\sqrt{t} - h_a(t) = 0$ and $0 \le P_0(t) \le 1$ there, and for $t \in [600a, 1]$ from the same facts and from (10.27) and (10.25).

Since here $a > 0$ is arbitrary, and $N_a \to \infty$ as $a \to 0$, (10.28) proves (10.11), and hence the second half of Theorem 10.1 has also been established, pending the construction of the polynomials P_0 with properties (10.26) and (10.27).

The rest of the proof consists of the construction of P_0 with properties (10.26) and (10.27). The proof has some common elements with the proof of Theorem 8.1, but there are some additional difficulties.

As before, let $a > 0$ be small, and consider the set

$$E_a = (E \cup [0,a] \cup [3a, 4a]) \setminus (a, 3a).$$

We know from Theorem 3.1 that for the equilibrium measure μ_{E_a} of E_a we have

$$\mu_{E_a}([0,a]) \geq M(a)\sqrt{a}$$

where $M(a) \to \infty$ as $a \to 0$. We may also assume that $\sqrt{a}M(a)$ is an increasing function of a. The representation (2.18) shows that for the Green function $g = g_{\mathbf{C}\setminus E_a}$ of $\mathbf{C} \setminus E_a$ we have for $\xi \in [a, 3a]$

$$g''(\xi) \leq -\int_0^a \frac{1}{(t-\xi)^2} d\mu_{E_a}(t) \leq -\frac{M(a)\sqrt{a}}{(3a)^2},$$

and since

$$-g\left(\frac{3}{2}a\right) + 2g(2a) - g\left(\frac{5}{2}a\right) = -\left(\frac{a}{2}\right)^2 g''(\xi) \geq \frac{M(a)\sqrt{a}}{36}$$

with some $\xi \in [a, 3a]$, we obtain

$$g_{\mathbf{C}\setminus E_a}(2a) \geq \frac{M(a)\sqrt{a}}{72}.$$

The value $g_{\mathbf{C}\setminus E_a}(2a)$ is the supremum of $|P_n(2a)|^{1/n}$ for all polynomials P_n of degree at most $n = 1, 2, \ldots$ with $|P_n(t)| \leq 1$ for $t \in E_a$ (see (2.21)), and since here we can square these polynomials, it follows that $g_{\mathbf{C}\setminus E_a}(2a)$ equals the above supremum even if P_n are required to be nonnegative on E. This gives that for all large $\nu \geq 1$ (not necessarily an integer) there are polynomials Q_ν of degree at most ν such that $Q_\nu(t) \geq 0$ for all t,

$$Q_\nu(t) \leq e^{-\nu M(a)\sqrt{a}/80} \qquad \text{for} \quad t \in E_a \tag{10.29}$$

and

$$\max_{t\in[a,3a]} Q_\nu(t) = 1.$$

Note that we have no control on $Q_\nu(t)$ for $t \in [3a, 1] \setminus E$. If it was small there then we could just integrate $Q_\nu(t)$ from t to 1 to get a polynomial that, after normalization, would be close to 1 on $[0, a]$ and would be small on $E_a \setminus [0, 3a]$. This integration process will work, if we multiply Q_ν by some polynomial R_{2j} of fixed degree and make sure that the integral of $Q_\nu R_{2j}$ is 0 on every interval contiguous to $E \setminus [0, 4a]$. Thus, let $[4a, 1] \setminus E = \cup_{i=1}^j I_i$, where the I_i's are disjoint intervals. We select two-two points $\xi_{i,1}, \xi_{i,2} \in I_i$ from each I_i, and consider the polynomial

$$R_{2j}(x) = \prod_{i=1}^j (x - \xi_{i,1})(x - \xi_{i,2}).$$

The number j depends on a, but for fixed a there is a constant K_a such that independently of the choice $\xi_{i,1}, \xi_{i,2} \in I_i$ we have $|R_{2j}(t)| \leq K_a$ for $t \in [0, 1]$ and $R_{2j}(t) \geq 1/K_a$ for $t \in [a, 3a]$. Now we claim

Lemma 10.6 *The numbers $\xi_{i,1}, \xi_{i,2} \in I_i$, $i = 1, \ldots, j$ can be chosen (depending on Q_ν) in such a way that we have*

$$\int_{I_i} Q_\nu(t) R_{2j}(t) dt = 0$$

for all $i = 1, \ldots, j$.

We shall prove the lemma at the end of the proof.

Set now

$$S_\nu^*(t) = \int_t^1 S_\nu(u) R_{2j}(u) du.$$

By the choice of R_{2j} it follows from (10.29) that

$$0 \le S_\nu^*(t) \le K_a e^{-\nu M(a)\sqrt{a}/80}$$

for $t \in E \cup [3a, 4a]$,

$$0 \le S_\nu^*(0) - S_\nu^*(t) = \int_0^t S_\nu(u) R_{2j}(u) du \le K_a e^{-\nu M(a)\sqrt{a}/80}$$

for $t \in [0, a]$, and

$$m_a := \max_{t \in [a, 3a]} S_\nu(t) R_{2j}(t) \ge \frac{1}{K_a}.$$

Let $t_0 \in [a, 3a]$ be a place where $S_\nu R_{2j}$ attains its maximum on $[a, 3a]$. It follows from Markov's inequality (7.1) that $S_\nu(t) R_{2j}(t) \ge m_a/2$ for $|t - t_0| \le a/2\nu^2$, which gives

$$S_\nu^*(0) \ge \frac{m_a}{2} \frac{a}{2\nu^2} \ge \frac{a}{4K_a \nu^2}.$$

Therefore, for large ν for the polynomials $S_\nu(t) = S_\nu^*(t)/S_\nu^*(0)$ of degree at most $\nu + 2j + 1$ we have $0 \le S_\nu(t) \le 1$ for $t \in [0, 4a]$,

$$0 \le 1 - S_\nu(t) \le e^{-\nu M(a)\sqrt{a}/100} \qquad \text{for} \quad t \in [0, a] \tag{10.30}$$

and

$$0 \le S_\nu(t) \le e^{-\nu M(a)\sqrt{a}/100} \qquad \text{for} \quad t \in E \cup [3a, 4a],\ t \ge 3a. \tag{10.31}$$

Recall, that this construction of S_ν depends on a, so let us write $S_{\nu,a}$ for S_ν to show this dependence.

We set for $t \ge a$

$$\varphi(t) = \sqrt{a} \log \frac{2t}{a}$$

and for a large n choose

$$\tau_{k,n} = \varphi^{-1}\left(\frac{2^k}{n}\right) = \frac{a}{2} \exp(2^k/\sqrt{a}n), \qquad \nu_{k,n} = \frac{2^{k+2} N_a}{\sqrt{\tau_{k,n}} M(\tau_{k,n})/100},$$

where these numbers are defined for those integer k for which

$$3\sqrt{a}n \le 2^k \le n\sqrt{a} \log \frac{2}{a} \tag{10.32}$$

(the N_a is the number from (10.23), and will be determined shortly). The $\nu_{k,n}$ are selected so that

$$\nu_{k,n} M(\tau_{k,n}) \sqrt{\tau}_{k,n}/100 = 2^{k+2} N_a$$

hence we obtain from (10.30) and (10.31)

$$0 \leq 1 - S_{\nu_{k,n}}(t) \leq e^{-2^{k+2} N_a} \qquad \text{for} \quad t \in [0, \tau_{k,n}] \tag{10.33}$$

and

$$0 \leq S_{\nu_{k,n}}(t) \leq e^{-2^{k+2} N_a} \qquad \text{for} \quad t \in E,\ t \geq 3\tau_{k,n}. \tag{10.34}$$

Let the set of the k's with property (10.32) be $\{k_0, k_0+1, \ldots, k_1\}$. It then follows that $e^3 a/2 \leq \tau_{k_0} \leq e^6 a/2 \leq 200a$. Consider the polynomial

$$P_0(t) = \prod_{k=k_0}^{k_1} S_{\nu_{k,n}, \tau_{k,n}}(t).$$

We claim that it satisfies (10.26) and (10.27). (10.26) is immediate from (10.33): for $t \in [0, a]$ we have

$$0 \leq 1 - P_0(t) \leq \sum_{k=k_0}^{k_1} e^{-2^{k+2} N_a} \leq n e^{-3\sqrt{a}n} \leq e^{-\sqrt{a}n}.$$

To prove (10.27) let $t \geq 600a$, $t \in E$. Then, since $\tau_{k_0} \leq 200a$. we have either $t \in [3\tau_{k,n}, 3\tau_{k+1,n}]$ for some $k = k_0, \ldots, k_1 - 1$, or $3\tau_{k_1} \leq t \leq 1$. In the former case we get from (10.34)

$$\begin{aligned} 0 \leq P_0(t) &\leq S_{\nu_{k,n}, \tau_{k,n}}(t) \leq \exp(-2^{k+2} N_a) \\ &\leq \exp(-2n N_a \varphi(\tau_{k+1,n})) \leq \exp(-2n N_a \varphi(t)), \end{aligned}$$

while in the latter case a similar argument with $k = 1$ gives the same inequality because

$$2^{k_1+2} \geq 2n\sqrt{a} \log \frac{2}{a} \geq 2n\varphi(t)$$

since k_1 is the largest k with property (10.32). Thus, (10.27) has been verified.

Finally, we have to estimate the degree of P_0. Exactly as in the proof of Theorem 8.1 we get for large n

$$\begin{aligned} \deg(P_0) &\leq \sum_{k=k_0}^{k_1} (\nu_{k,n} + 2j + 1) \leq \sum_{k=k_0}^{k_1} 2\nu_{k,n} \\ &= \sum_{k=k_0}^{k_1} \frac{2 \cdot 2^{k+2} N_a}{\sqrt{\varphi^{-1}(2^k/n)} M(\varphi^{-1}(2^k/n))/100} \\ &\leq 1600 n N_a \int_{3\sqrt{a}/2}^{\sqrt{a} \log 2/a} \frac{du}{\sqrt{\varphi^{-1}(u)} M(\varphi^{-1}(u))}. \end{aligned}$$

In the last integral we have

$$\varphi^{-1}(u) = \frac{a}{2} e^{u/\sqrt{a}},$$

therefore the integral equals

$$\int_{3\sqrt{a}/2}^{\sqrt{a}\log 2/a} \frac{du}{\sqrt{a/2}e^{u/2\sqrt{a}}M((a/2)e^{u/\sqrt{a}})},$$

and the change of variable $v = e^{u/2\sqrt{a}}$ changes this into

$$\frac{1}{\sqrt{a/2}} \int_{e^{3/4}}^{\sqrt{2/a}} \frac{1}{v}\frac{2\sqrt{a}}{v}\frac{1}{M(av^2/2)}dv =: d_a$$

where $d_a \to 0$ as $a \to 0$ because $M(a) \to \infty$ as $a \to 0$. Thus, the degree of P_0 is at most $1600nN_a d_a \leq n$ provided N_a tends to infinity as $a \to 0$ so slowly that we have $1600N_a d_a \leq 1$ for sufficiently small a.

■

Proof of Lemma 10.6. Let $I_i = (\alpha_i, \beta_i)$, and for $t_i \in [-1, 1]$ set

$$\xi_{i,1} = \frac{\alpha_i + \beta_i}{2} - \frac{1 + t_i}{2}\frac{\beta_i - \alpha_i}{2},$$

$$\xi_{i,2} = \frac{\alpha_i + \beta_i}{2} + \frac{1 + t_i}{2}\frac{\beta_i - \alpha_i}{2}.$$

Then for $t_i = -1$ we have $\xi_{i,1} = \xi_{i,2} = (\alpha_i + \beta_i)/2$, so independently of the other t_s's

$$\int_{I_i} Q_\nu(t)R_{2j}(t)dt > 0.$$

On the other hand, if $t_i = 1$ then $\xi_{i,1} = \alpha_i$, $\xi_{i,2} = \beta_i$, so independently of the other t_s's

$$\int_{I_i} Q_\nu(t)R_{2j}(t)dt < 0.$$

Thus, the mapping

$$\underline{\mathbf{t}} = (t_1, \ldots, t_j) \to \mathbf{Y}(\underline{\mathbf{t}}) = (y_1(t_1, \ldots, t_j), \ldots, y_j(t_1, \ldots, t_j)),$$

$$y_i(t_1, \ldots, t_j) = \int_{I_i} Q_\nu(t)R_{2j}(t)dt$$

is a mapping from $[-1, 1]^j$ into $\mathbf{R}^j$ with the property that the face $t_i = -1$ of the cube $[-1, 1]^j$ is mapped into the positive half space $y_i > 0$, while the face $t_i = 1$ of the cube is mapped into the negative half space $y_i < 0$. Such a mapping contains the origin of $\mathbf{R}^j$ in the image set, and this is exactly the statement of the lemma. Indeed, this can be proved by the same method that we used in the proof of Proposition 6.6: if $\underline{\mathbf{0}}$ was not in the image set, then let $P(\underline{\mathbf{t}})$ be the point where the half-line emanating from the origin and going through the point $\mathbf{Y}(\underline{\mathbf{t}})$ hits the boundary of the cube $[-1, 1]^j$. From what we have said it follows that this is a continuous mapping from $[-1, 1]^j$ into its boundary without fixed point, and since this contradicts the Brouwer fixed point theorem, the lemma follows.

■

Proof of Theorem 10.3. We follow the proof of Theorem 10.2, and first let us prove the sufficiency of the condition $\sum \varepsilon_i^2 < \infty$, that corresponds to the first part of Theorem 10.2. That proof was based on two things. First of all it was based on the Markov inequality (7.16), the analogue of which in the present case is (7.18). The second ingredient was the existence of the points $x_n \in [2^{-2n}, 2^{-2n+1)}]$, $x_n, 2x_n, 3x_n, \ldots, (m+1)x_n \in E$ and the choice of the expression

$$\Delta_{n,m} f = x_n^{-m} \sum_{i=0}^{m} (-1)^i \binom{m}{i} f((m+1-i)x_n)$$

for which we used the properties:

$$\Delta_{n,m} f = f^{(m)}(\xi_n)$$

and

$$|\Delta_{n,m} f| \leq 2^m x_n^{-m} \|f\|_E.$$

Now this choice may fail for $E = \mathcal{C}(\varepsilon_1, \varepsilon_2, \ldots)$, since the latter set may be of measure 0, and we cannot use the density argument to guarantee the existence of an x_n such that all the points $x_n, 2x_n, 3x_n, \ldots, (m+1)x_n$ belong to E. However, it is clear that the above choice can be replace by a choice of $m+1$ points $x_{n0} < \ldots < x_{nm}$ from E and by a choice of

$$\tilde{\Delta}_{n,m} f = a_{n0} f(x_{n0}) + \cdots + a_{nm} f(x_{nm}),$$

provided we have the properties: there are constants L and $\beta < \alpha < 1$ such that

$$x_{nm} \leq L x_{n0}, \qquad \beta x_{n,0} \leq x_{n+1,0} \leq \alpha x_{n,0} \tag{10.35}$$

$$\tilde{\Delta}_{n,m} f = f^{(m)}(\xi), \qquad \text{for some } \xi \in [x_{n0}, x_{nm}] \tag{10.36}$$

and

$$|\Delta_{n,m} f| \leq L x_{n0}^{-m} \|f\|_E. \tag{10.37}$$

In fact, then instead of x_n we can use that $x_{n'0}$ for which $x_{n'0} \in [2^{-2n}, 2^{-2(n-1)}]$ and instead of $\Delta_{n+s,m}$ we can use $\tilde{\Delta}_{n'+s,m} f$; and the proof goes over with little changes.

Let now

$$t_n = \frac{1-\varepsilon_1}{2} \frac{1-\varepsilon_2}{2} \cdots \frac{1-\varepsilon_n}{2}$$

be the numbers from (5.3). It is clear from the Cantor construction that the points $x_{ni} = t_{n+m+1-i}$ belong to $\mathcal{C}(\varepsilon_1, \varepsilon_2, \ldots)$ and they satisfy (10.35). Instead of forward difference now we have to use a constant multiple of divided differences:

$$\tilde{\Delta}_{n,m} f = m! \sum_{i=0}^{m} \frac{1}{\prod_{j \neq i} (x_{n,i} - x_{n,j})} f(x_{ni}),$$

for which it is well known (see e.g. [17, (4.7.4), p. 120]) that (10.36) is true. Finally, since in our case we have $x_{n,i+1} - x_{n,i} \geq \gamma x_{n0}$ with some $\gamma > 0$, property (10.37) is also true.

With this the sufficiency of the condition $\sum_i \varepsilon_i^2 < \infty$ has been verified.

Next assume that $\sum_i \varepsilon_i^2 = \infty$, and consider the set

$$E = \tilde{C}(\varepsilon_1, \varepsilon_2, \dots) = [0,1] \setminus \left(\bigcup_{k=1}^{\infty} (t_k, t_k + \varepsilon_k t_{k-1}) \right)$$

from (5.2). As in the proof of Theorem 10.1, let $a > 0$ be small, and consider the set

$$E_a = (E \cup [0,a] \cup [3a, 4a]) \setminus (a, 3a).$$

It follows from Corollary 3.2 (c.f. also Corollary 3.3 and its proof as well as the beginning of the proof of Theorem 5.1) that $\sum_i \varepsilon_i^2 = \infty$ implies

$$g_{\mathbf{C}\setminus E_a}(-a) \geq M_1(a)\sqrt{a}$$

where $M_1(a) \to \infty$ as $a \to 0$. But this gives via Harnack's inequality that

$$g_{\mathbf{C}\setminus E_a}(a) \geq M(a)\sqrt{a} \tag{10.38}$$

is true with some $M(a) \to \infty$ as $a \to 0$. Now we can copy the proof of the second part of Theorem 10.1 with this E. In fact, of E there two properties were used, namely (10.38) and the property that every set $[b,1] \setminus E$, $b > 0$ consists of finitely many intervals, and both conditions are true in the present case.

Thus, by that proof we obtain that

$$\lim_{n\to\infty} n^{2p} \mathcal{E}_n(|x|^p, E) = 0,$$

and since $\mathbf{C}(\varepsilon_1, \varepsilon_2, \dots) \subset E$, the necessity of the condition $\sum_i \varepsilon_i^2 < \infty$ has also been verified.

■

Proof of Theorem 10.5. The outline of the proof is the following: we shall use Bernstein's result (10.2) for $[-1,1]$, and transform it to special sets E of the form $T_N^{-1}([-1,1])$ where T_N is a polynomial with some special properties. The point is that by applying polynomial inverse images the equilibrium measures are preserved. The next step is to approximate arbitrary sets E consisting of finitely many intervals by such polynomial inverse images $T_N^{-1}([-1,1])$. The final step is to approximate an arbitrary compact by sets consisting of finitely many intervals. This approach to transform results from one interval to general sets has been applied for polynomial inequalities in [38].

The polynomials T_N with respect to which we take the inverse image of $[-1,1]$ have to satisfy the following condition: if the degree of T_N is N, then T_N has $N-1$ local extrema, and the local extremal values are alternately ≥ 1 and ≤ -1. In this case we say that T_N is admissible. It implies that T_N has N real zeros, and as x runs trough the real line, the value $T_N(x)$ runs through (among others) the interval $[-1,1]$ N times. Thus, there are precisely N inverse branches of $T_N^{-1}([-1,1])$, i.e. there are N intervals that are mapped by T_N onto $[-1,1]$. Some of these intervals however may combine into a single intervals, hence eventually the inverse image may consist of some number of intervals, where this number can be anything from 1 to N. E.g. if T_N is the classical Chebyshev polynomial of $[-1,1]$ (normalized to have norm 1), then all the N intervals combine into a single one, and the inverse image of $[-1,1]$ is just one interval ($[-1,1]$).

As we have just explained, first we prove the claim for sets of the special form: $E = T_N^{-1}([-1,1])$, where T_N is an admissible polynomial, and suppose that x_0 is a point in E with property $T_N(x_0) = 0$. It is well known [19, Theorem 11], [29] that

$$\omega_E(x_0) = \frac{|T_N'(x_0)|}{\pi N\sqrt{1 - T_N(x_0)^2}}, \tag{10.39}$$

which takes the form

$$\omega_E(x_0) = \frac{|T_N'(x_0)|}{\pi N} \tag{10.40}$$

in the present case.

Let $\delta > 0$ be so small that on the interval $(x_0 - 2\delta, x_0 + 2\delta)$ the polynomial T_N is monotone (it is monotone on the N inverse subintervals of $[-1,1]$, and x_0 is in the interior of one of them). Let $E \subset [-A, A]$, and select a function $\psi \in C^{4p}[-A, A]$ (i.e. ψ is $[4p]$ times continuously differentiable and $\psi^{[4p]} \in \mathrm{Lip}\,(4p - [4p])$), that equals 1 on $(x_0 - \delta, x_0 + \delta)$ and equals zero outside $(x_0 - 2\delta, x_0 + 2\delta)$. The functions $\psi(x)$ and

$$\varphi(x) = \psi(x)|T_N(x)|^p - |x - x_0|^p|T_N'(x_0)|^p$$

are both C^{4p} functions on $[-A, A]$, therefore by Jackson's theorem they can be approximated by an error Cm^{-4p} by polynomials of degree at most m. Thus, there are polynomials $Q_{\sqrt{n}}$ and $R_{\sqrt{n}}$ of degree at most $\sqrt{n}$ such that uniformly on $[-A, A]$

$$|\psi(x) - Q_{\sqrt{n}}(x)| \le \frac{C}{n^{2p}} \tag{10.41}$$

$$|\varphi(x) - R_{\sqrt{n}}(x)| \le \frac{C}{n^{2p}}.$$

This gives uniformly on $[-A, A]$

$$\Big|Q_{\sqrt{n}}(x)|T_N(x)|^p - |x - x_0|^p|T_N'(x_0)|^p - R_{\sqrt{n}}(x)\Big| \le \frac{C}{n^{2p}}$$

with a C that may depend on T_N. On the other hand, Bernstein's theorem (10.2) gives that for any $\varepsilon > 0$ and sufficiently large n there are polynomials P_n of degree at most n such that

$$\Big||x|^p - P_n(x)\Big| \le (1+\varepsilon)\frac{\sigma_p}{n^p}, \qquad x \in [-1,1],$$

which is the same as

$$\Big||T_N(x)|^p - P_n(T_N(x))\Big| \le (1+\varepsilon)\frac{\sigma_p}{n^p}, \qquad x \in E.$$

Thus, we obtain from

$$|Q_{\sqrt{n}}(x)| \le \psi(x) + \frac{C}{n^{2p}} \le 1 + \frac{C}{n^{2p}}, \qquad x \in [-A, A]$$

the estimate

$$\Big|Q_{\sqrt{n}}(x)P_n(T_N(x)) - |x - x_0|^p|T_N'(x_0)|^p - R_{\sqrt{n}}(x)\Big| \le (1+\varepsilon)\frac{\sigma_p}{n^p} + \frac{C}{n^{2p}}.$$

This shows that

$$\mathcal{E}_{nN+\sqrt{n}}(|x-x_0|^p, E) \le (1+\varepsilon)\frac{\sigma_p}{|T_N'(x_0)|^p n^p} + \frac{C}{|T_N'(x_0)|^p n^{2p}},$$

and so for $n \to \infty$ we obtain

$$\limsup_{m\to\infty} m^p \mathcal{E}_m(|x-x_0|^p, E) \le (1+\varepsilon)\frac{\sigma_p N^p}{|T_N'(x_0)|^p} = (1+\varepsilon)(\pi\omega_E(x_0))^{-p}\sigma_p,$$

where at the last step we used (10.40). Since here $\varepsilon > 0$ is arbitrary, this proves the appropriate upper estimate for $\mathcal{E}_n$ in (10.18).

Conversely, let us suppose that for some σ^* we have

$$\liminf_{n\to\infty} n^p \mathcal{E}_n(|x-x_0|^p, E) \le \sigma^*.$$

Then the lower estimate in (10.18) is the same as the inequality

$$\sigma^* \ge \sigma_p(\pi\varepsilon_E(x_0))^{-p},$$

which we are going to prove below.

Thus, suppose that for all $\varepsilon > 0$ and for infinitely many n, say $n \in \mathcal{N}$, there are polynomials P_n of degree n with the property

$$\Big||x-x_0|^p - P_n(x)\Big| \le (1+\varepsilon)\frac{\sigma^*}{n^p}, \qquad x \in E.$$

On multiplying through by $\psi(x)$ and on applying (10.41) we arrive at

$$\Big|\psi(x)|x-x_0|^p - P_n(x)Q_{\sqrt{n}}(x)\Big| \le (1+\varepsilon)\frac{\sigma^*}{n^p} + \frac{C}{n^{2p}}, \qquad x \in E, \tag{10.42}$$

where $Q_{\sqrt{n}}$ are the polynomials from (10.41). Since the function

$$\psi(x)\frac{|T_N(x)|^p}{|T_N'(x_0)|^p} - \psi(x)|x-x_0|^p$$

is in C^{4p}, we may approximate it in the order C/n^{2p} by polynomials of degree at most $\sqrt{n}$, hence there are polynomials $R^*_{\sqrt{n}}$ of degree at most $\sqrt{n}$ such that

$$\left|\psi(x)\frac{|T_N(x)|^p}{|T_N'(x_0)|^p} - \psi(x)|x-x_0|^p - R^*_{\sqrt{n}}(x)\right| \le \frac{C}{n^{2p}}, \qquad x \in [-A, A]. \tag{10.43}$$

This and (10.42) give

$$\left|\psi(x)\frac{|T_N(x)|^p}{|T_N'(x_0)|^p} - R^*_{\sqrt{n}}(x) - P_n(x)Q_{\sqrt{n}}(x)\right| \le (1+\varepsilon)\frac{\sigma^*}{n^p} + \frac{C}{n^{2p}}, \qquad x \in E. \tag{10.44}$$

For $x \in E$ let $x_1, \ldots, x_N$ be the N values of $T_N^{-1}(T_N(x))$, and consider the expression

$$S_{n+\sqrt{n}}(x) = \sum_{i=1}^{N}(R^*_{\sqrt{n}}(x_i) + P_n(x_i)Q_{\sqrt{n}}(x_i)) \tag{10.45}$$

This is a symmetric polynomial in $x_1, \dots, x_n$, so it can be written as a polynomial of the elementary symmetric polynomials of these variables. However, $x_1, \dots, x_n$ are the zeros of the equation (in t) $T_N(t) - T_N(x) = 0$, therefore if

$$T_N(t) = \sum_{j=0}^{N} b_j t^j,$$

then the value of the elementary symmetric polynomials in $x_1, \dots, x_n$ are nothing else than $\pm b_j/b_N$, except for the last one, which is equal to $(-1)^N(b_0 - T_N(x))/b_N$. Thus, $S_{n+\sqrt{n}}(x)$ is actually a polynomial in $T_N(x)$ of degree at most $(n+\sqrt{n})/N$, that is $S_{n+\sqrt{n}}(x) = S^*_{(n+\sqrt{n})/N}(T_N(x))$ with a polynomial $S^*_{(n+\sqrt{n})/N}$ of degree at most $(n+\sqrt{n})/N$. Since ψ vanishes outside $(x_0 - 2\delta, x_0 + 2\delta)$, and at most one x_i belongs to this interval, it follows from (10.41) and (10.43) that at most one term in the sum (10.45) is larger than C/n^{2p}. This and (10.44) give

$$\left|\sum_{i=1}^{N} \psi(x_i)\frac{|T_N(x_i)|^p}{|T_N'(x_0)|^p} - S^*_{(n+\sqrt{n})/N}(T_N(x))\right| \le (1+\varepsilon)\frac{\sigma^*}{n^p} + \frac{CN}{n^{2p}}.$$

On the left we have

$$\sum_{i=1}^{N} \psi(x_i)\frac{|T_N(x_i)|^p}{|T_N'(x_0)|^p} = \frac{|T_N(x)|^p}{|T_N'(x_0)|^p}\sum_{i=1}^{N}\psi(x_i) = \frac{|T_N(x)|^p}{|T_N'(x_0)|^p}\Psi(T_N(x))$$

where for $u \in [-1,1]$ the function $\Psi(u)$ is defined as

$$\Psi(u) = \sum_{T_N(u_i)=u} \psi(u_i).$$

Summing up, we obtain that for all $n \in \mathcal{N}$

$$\left|\Psi(T_N(x))\frac{|T_N(x)|^p}{|T_N'(x_0)|^p} - S^*_{(n+\sqrt{n})/N}(T_N(x))\right| \le (1+\varepsilon)\frac{\sigma^*}{n^p} + \frac{CN}{n^{2p}}, \qquad x \in E,$$

which gives

$$\mathcal{E}_{(n+\sqrt{n})/N}\left(\Psi(u)\frac{|u|^p}{|T_N'(x_0)|^p}, [-1,1]\right) \le (1+\varepsilon)\frac{\sigma^*}{n^p} + \frac{CN}{n^{2p}}.$$

This is valid for all $n \in \mathcal{N}$, hence

$$\liminf_{n\to\infty} m^p E_m\left(\Psi(u)|u|^p, [-1,1]\right) \le \sigma^*\frac{|T_N'(x_0)|^p}{N^p} = \sigma^*(\pi\omega_E(x_0))^p.$$

Since $\Psi(u) = 1$ in a neighborhood of the origin, the function $|u|^p - \Psi(u)|u|^p$ is in C^{4p} and so it can be approximated in the order n^{-2p} by polynomials of degree at most $\sqrt{n}$, hence the preceding relation gives also

$$\liminf_{n\to\infty} m^p E_m\left(|u|^p, [-1,1]\right) \le \sigma^*(\pi\omega_E(x_0))^p.$$

On comparing this with Bernstein's result (10.2) we obtain

$$\sigma^* \ge \sigma_p(\pi\omega_E(x_0))^{-p},$$

and this proves the appropriate lower estimate in (10.18).

Next we prove (10.18) for sets E that consist of finitely many intervals: $E = \cup_{i=1}^{l}[a_i, b_i]$, where $a_1 < b_1 < a_2 < \cdots < a_l < b_l$. We will prove this case by approximating E by sets of the form $T_N^{-1}([-1,1])$ with some appropriate admissible T_N, and therefore reducing the problem to the case discussed and proved above. To this end we need

Lemma 10.7 *Let $x_0 \in E$ be in the interior of $E = \cup_{i=1}^{l}[a_i, b_i]$, $a_1 < b_1 < a_2 < \cdots < a_l < b_l$. For every $\varepsilon > 0$ there is an admissible polynomial T_N such that $T_N(x_0) = 0$, $E^* = T_N^{-1}([-1,1])$ consists of l intervals: $E^* = \cup_{i=1}^{l}[a_i^*, b_i^*]$, where $a_1^* < b_1^* < a_2^* < \cdots < a_l^* < b_l^*$, and for each i we have*

$$|a_i - a_i^*| \le \varepsilon, \qquad |b_i - b_i^*| \le \varepsilon. \tag{10.46}$$

Besides that we may also request that $E^* \subset E$, or alternatively $E \subset E^*$, be satisfied.

Proof. Without the requirement $T_N(x_0) = 0$ this is Theorem [38, Theorem 2.1] (see also [16] and [28]). From it we can get the claim as follows. As we have just said, there is an admissible polynomial S_M such that, $E^\# = S_M^{-1}([-1,1])$ consists of l intervals: $E^\# = \cup_{i=1}^{l}[a_i^\#, b_i^\#]$, where $a_1^\# < b_1^\# < a_2^\# < \cdots < a_l^\# < b_l^\#$, and for each i we have $|a_i - a_i^\#| < \varepsilon/2$, $|b_i - b_i^\#| < \varepsilon/2$. Let $\mathcal{T}_m(x) = \cos(m \arccos x)$ be the Chebyshev polynomials of $[-1,1]$. Note that $\mathcal{T}_m(S_M)$ is again an admissible polynomial with the same inverse set as for S_M:

$$E^\# = S_M^{-1}([-1,1]) = (\mathcal{T}_m(S_M))^{-1}([-1,1]).$$

Since for any $y \in [-1,1]$ and any $\delta > 0$ there is a zero of $\mathcal{T}_m$ in $(y-\delta, y+\delta)$ if m is sufficiently large, it follows that for large m there is a zero x_1 of $\mathcal{T}_m(S_M)$ in $(x_0 - \varepsilon/2, x_0 + \varepsilon/2)$. Now the polynomial $T_N(x) = \mathcal{T}_m(S_M(x + x_1 - x_0))$ is again admissible, $T_N(x_0) = 0$, and since $T_N^{-1}([-1,1]) = E^\# + x_0 - x_1$, we have the relations (10.46) for it.

The additional property $E^* \subseteq E$ (or $E \subseteq E^*$) can automatically be achieved: just apply what has been proven to the set $\tilde{E} = \cup_{i=1}^{l}[a_i + \varepsilon/2, b_i - \varepsilon_i/2]$ and to the number $\varepsilon/2$.

■

Returning now to the proof of Theorem 10.5, let E consist of finitely many intervals, $E = \cup_{i=1}^{l}[a_i, b_i]$, $a_1 < b_1 < a_2 < \cdots < a_l < b_l$. For a given $x_0 \in \mathrm{Int}(E)$ and $\varepsilon > 0$ choose the set E^* as in the lemma, with the additional property that $E^* \subset E$. As we have already proved the result for the set E^* and the point x_0, we obtain

$$\liminf_{n\to\infty} n^p \mathcal{E}_n(|x - x_0|^p, E) \ge \liminf_{n\to\infty} n^p \mathcal{E}_n(|x - x_0|^p, E^*) = (\pi\omega_{E^*}(x_0))^{-p}\sigma_p,$$

and it easily follows from the explicit form of ω_E given in (10.17) (or apply Lemma 10.8 below) that we have $\omega_{E^*}(x_0) \to \omega_E(x_0)$ as $\varepsilon \to 0$. This gives

$$\liminf_{n\to\infty} n^p \mathcal{E}_n(|x - x_0|^p, E) \ge (\pi\omega_E(x_0))^{-p}\sigma_p. \tag{10.47}$$

In a similar fashion, if now E^* is chosen as in the lemma, but with the additional property that $E \subset E^*$, then we obtain

$$\limsup_{n\to\infty} n^p \mathcal{E}_n(|x-x_0|^p, E) \le \limsup_{n\to\infty} n^p \mathcal{E}_n(|x-x_0|^p, E^*) = (\pi\omega_{E^*}(x_0))^{-p}\sigma_p,$$

and from this

$$\limsup_{n\to\infty} n^p \mathcal{E}_n(|x-x_0|^p, E) \le (\pi\omega_E(x_0))^{-p}\sigma_p \tag{10.48}$$

follows exactly as before. (10.47) and (10.48) prove the result when E consists of finitely many intervals.

Finally we deal with the case of general compact set E. Thus, let $E \subset \mathbf{R}$ be an arbitrary compact set and $x_0 \in \mathrm{Int}(E)$. Let $\delta > 0$ be so small that $(x_0-3\delta, x_0+3\delta) \subset E$.

(10.48) follows fairly directly: by the outer regularity of logarithmic capacity, for every $\varepsilon > 0$ there is an open set $E \subset U_\varepsilon$ such that $\mathrm{cap}(U_\varepsilon) \le \mathrm{cap}(E) + \varepsilon$. Since E is compact, we may assume U_ε to consist of finitely many intervals, and then we may apply the finitely many interval case to the closure $\overline{U_\varepsilon}$ to deduce

$$\limsup_{n\to\infty} n^p \mathcal{E}_n(|x-x_0|^p, E) \le \limsup_{n\to\infty} n^p \mathcal{E}_n(|x-x_0|^p, \overline{U_\varepsilon}) = (\pi\omega_{\overline{U_\varepsilon}}(x_0))^{-p}\sigma_p. \tag{10.49}$$

Now we use the following lemma that we prove after completing the proof of Theorem 10.5.

Lemma 10.8 *Let F be compact, $I \subset F$ an open interval, and suppose that $F \subseteq F_k$ are compact sets lying in a fixed interval such that $\mathrm{cap}(F_k) \to \mathrm{cap}(F)$ as $k \to \infty$. Then*

$$\lim_{k\to\infty} \omega_{F_k}(t) = \omega_F(t) \tag{10.50}$$

uniformly on compact subsets of I.

The same conclusion holds if instead of $F \subseteq F_k$ we assume $I \subset F_k \subseteq F$ for all k.

Recall that $\omega_E(x)$ is the density (Radon-Nikodym derivative) of the equilibrium measure μ_E (see (10.17).

By this lemma if $\varepsilon > 0$ is small, the value $\omega_{\overline{U_\varepsilon}}(x_0)$ is close to $\omega_E(x_0)$ and so (10.49) proves (10.48).

To prove (10.47) choose $\delta > 0$ so that $[x_0 - 4\delta, x_0 + 4\delta] \subseteq E$. Let $h_{x_0}(x)$ be a function such that $h_{x_0}(x)$ coincides with $|x-x_0|^p$ on $[x_0-\delta/2, x_0+\delta/2]$, $h_{x_0}(x) = 0$ outside $[x_0-\delta, x_0+\delta]$ and $|x-x_0|^p - h_{x_0}(x) \in C^{4p}$. Then the latter function can be approximated on any interval in the order n^{-4p} by polynomials of order n, hence

$$\liminf_{n\to\infty} n^p \mathcal{E}_n(|x-x_0|^p, E) = \liminf_{n\to\infty} n^p \mathcal{E}_n(h_{x_0}, E),$$

and so in what follows we may work with h_{x_0} instead of $|x-x_0|^p$.

Without loss of generally assume $E \subset [-1,1]$ (apply linear transformation if this is not the case).

Choose a small $\varepsilon > 0$. It is well known (see e.g. [30, Corollary VI.3.6]) that there is an $\eta = \eta(\varepsilon) > 0$ such that for all large n there are polynomials $Q_{\varepsilon n}$ of degree at most εn such that $0 \le Q_{\varepsilon n}(x) \le 1$ for all $x \in [-1,1]$, $|Q_{\varepsilon n}(x) - 1| \le e^{-\eta\varepsilon n}$ for $x \in [x_0-\delta, x_0+\delta]$, and $|Q_{\varepsilon n}(x)| \le e^{-\eta\varepsilon n}$ for $x \in [-1,1] \setminus [x_0-2\delta, x_0+2\delta]$.

Let $g_{\mathbf{C}\setminus(E\setminus[x_0-2\delta,x_0+2\delta])}(z)$ be the Green function of the set $\mathbf{C}\setminus(E\setminus(x_0-2\delta,x_0+2\delta))$ with pole at infinity, and for a given $\varepsilon>0$ let $V_{\eta\varepsilon/2}$ be the level set

$$V_{\eta\varepsilon/2}=\{z\,|\,g_{\mathbf{C}\setminus(E\setminus(x_0-2\delta,x_0+2\delta))}(z)<\eta\varepsilon/2\}.$$

Since Green functions are subharmonic (and hence they are upper semi-continuous), it follows that $V_{\eta\varepsilon/2}$ is open, and so $(E\setminus(x_0-2\delta,x_0+2\delta))\setminus V_{\eta\varepsilon/2}$ is closed. Since every regular point (with respect to the Dirichlet problem in $\mathbf{C}\setminus(E\setminus(x_0-2\delta,x_0+2\delta)))$ of $E\setminus(x_0-2\delta,x_0+2\delta)$ is contained in $V_{\eta\varepsilon/2}$, the set $(E\setminus(x_0-2\delta,x_0+2\delta))\setminus V_{\eta\varepsilon/2}$ is part of the set of irregular points, hence it is of capacity 0. By Lemma [30, VI.2.2] there is a $\kappa=\kappa(\varepsilon)>0$ such that for any compact set $K\subset[-1,1]\setminus[x_0-2\delta,x_0+2\delta]$ there are for all sufficiently large n polynomials $R_{\varepsilon n}$ of degree at most $\varepsilon>0$ such that $|R_{\varepsilon n}(x)|\le 2$ for all $x\in[-1,1]$, $|R_{\varepsilon n}(x)-1|\le 2^{-\kappa\varepsilon n}$ for $x\in[x_0-\delta,x_0+\delta]$ and $|R_{\varepsilon n}(x)|\le \mathrm{cap}(K)^{\kappa\varepsilon n}$ for $x\in K$. If we consider $R_{\varepsilon n}^2(x)(4-R_{\varepsilon n}^2(x))^3/27$ instead of $R_{\varepsilon n}(x)$, we may even assume that $0\le R_{\varepsilon n}(x)\le 1$ for all $x\in[-1,1]$ (note that the function $x\to x^2(4-x^2)^3/27$ maps the interval $[-2,2]$ into $[0,1]$ and leaves 0 and 1 invariant). Choose an open set $K=K_{\eta,\varepsilon,\delta}$ such that $(E\setminus(x_0-2\delta,x_0+2\delta))\setminus V_{\eta\varepsilon/2}\subset K$, K consists of finitely may intervals and $\mathrm{cap}(\overline{K})^{\kappa\varepsilon}<\delta$. Thus, the $R_{\varepsilon n}(x)$ for this K satisfies $|R_{\varepsilon n}(x)|\le\delta^n$ for $x\in K$.

Now consider the set $(x_0-2\delta,x_0+2\delta)\cup K\cup V_{\eta\varepsilon/2}$, which is an open cover of the compact set E. Hence there are open sets $K'\subset K$ and $V'_{\eta\varepsilon/2}\subset V_{\eta\varepsilon/2}$ consisting of finitely many intervals such that $E\subset H:=(x_0-2\delta,x_0+2\delta)\cup K'\cup V'_{\eta\varepsilon/2}$. Let P_n be the best approximation of h_{x_0} on E, and consider the polynomials $S_{(1+2\varepsilon)n}=P_nQ_{\varepsilon n}R_{\varepsilon n}$ of degree at most $(1+2\varepsilon)$. Note that since h_{x_0} vanishes outside $(x_0-\delta,x_0+\delta)$, we have

$$|S_{(1+2\varepsilon)n}(x)|\le\mathcal{E}_n(h_{x_0},E)Q_{\varepsilon n}(x)R_{\varepsilon n}(x)\quad x\in E\setminus[x_0-\delta,x_0+\delta].$$

Now since for $x\in[x_0-\delta,x_0+\delta]$ we have (by the choice of these polynomials)

$$1-e^{-\eta\varepsilon n}\le Q_{\varepsilon n}(x)\le 1,\qquad 1-2^{-\kappa\varepsilon n}\le R_{\varepsilon n}(x)\le 1,$$

we can write

$$|S_{(1+2\varepsilon)n}(x)-h_{x_0}(x)|\le\mathcal{E}_n(h_{x_0},E)+4\cdot e^{-\kappa\varepsilon n}+4\cdot e^{-\eta\varepsilon n},\quad x\in[x_0-\delta,x_0+\delta]$$

and from

$$h_{x_0}(x)=0,\qquad 0\le Q_{\varepsilon n}(x)\le 1,\qquad R_{\varepsilon n}(x)\le 1$$

for $x\in[x_0-2\delta,x_0-\delta]\cup[x_0+\delta,x_0+2\delta]$ we can conclude

$$|S_{(1+2\varepsilon)n}(x)|\le\mathcal{E}_n(h_{x_0},E),\quad x\in[x_0-2\delta,x_0-\delta]\cup[x_0+\delta,x_0+2\delta].$$

By the Bernstein-Walsh lemma (2.20) we get for $x\in(V'_{\eta\varepsilon/2}\cap[-1,1])\setminus(x_0-2\delta,x_0+2\delta)$ from

$$0\le Q_{\varepsilon n}(x)\le e^{-\eta\varepsilon n},\qquad g_{\mathbf{C}\setminus(E\setminus(x_0-2\delta,x_0+2\delta))}(z)<\eta\varepsilon/2$$

the estimate

$$\begin{aligned}|S_{(1+2\varepsilon)n}(x)| &\le \|P_n\|_E e^{ng_{\mathbf{C}\setminus(E\setminus(x_0-2\delta,x_0+2\delta))}(x)}e^{-\eta\varepsilon n}\\ &\le \|P_n\|_E e^{\eta\varepsilon n/2}e^{-\eta\varepsilon n}=\|P_n\|_E e^{-\eta\varepsilon n/2}.\end{aligned}$$

In a similar manner, for $x \in (K' \cap [-1,1]) \setminus (x_0 - 2\delta, x_0 + 2\delta)$ we can use

$$0 \le Q_{\varepsilon n}(x) \le e^{-\eta \varepsilon n}, \qquad 0 \le R_{\varepsilon n}(x)| \le \delta^n$$

and

$$\|P_n\|_{[0,1]} \le \left(\frac{4}{8\delta}\right)^n \|P_n\|_E \le \delta^{-n}\|P_n\|_E$$

which follows from (2.24) because $(x_0 - 4\delta, x_0 + 4\delta) \subseteq E$. These give for $x \in (K' \cap [-1,1]) \setminus (x_0 - 2\delta, x_0 + 2\delta)$

$$|S_{(1+2\varepsilon)n}(x)| \le \|P_n\|_E \delta^{-n}\delta^n e^{-\eta\varepsilon n} = \|P_n\|_E e^{-\eta\varepsilon n}.$$

Thus,

$$\|h_{x_0} - S_{(1+2\varepsilon)n}\|_{\overline{H}} \le \mathcal{E}_n(h_{x_0}, E) + o(n^{-p}),$$

and hence

$$\begin{aligned} \liminf_{n\to\infty} n^p \mathcal{E}_n(h_{x_0}, E) &\ge \liminf_{n\to\infty} n^p \mathcal{E}_{(1+2\varepsilon)n}(h_{x_0}, \overline{H}) \\ &= (1+2\varepsilon)^{-p}(\pi\omega_{\overline{H}}(x_0))^{-p}\sigma_p \\ &\ge (1+2\varepsilon)^{-p}(\pi\omega_E(x_0))^{-p}\sigma_p, \end{aligned}$$

where at the equality we have used that $\overline{H}$ consists of finitely many intervals and so for it we already know (10.18) (use also that $\mathcal{E}_{(1+2\varepsilon)n}(h_{x_0}, \overline{H})$ and $\mathcal{E}_{(1+2\varepsilon)n}(|x - x_0|^p, \overline{H})$ differ by at most Cn^{-4p}). The last inequality is a consequence of $E \subseteq \overline{H}$.

Now this proves (10.47) and the proof of (10.18) is complete, pending the proof of Lemma 10.8.

∎

Proof of Lemma 10.8. First we show that the equilibrium measures μ_{F_k} converge in the weak* topology to μ_F. In fact, let σ be a weak* limit of some subsequence, say $\mu_{F_{k_l}} \to \sigma$ as $l \to \infty$ and let J be an interval containing the support of all F_k. Then σ is supported in J, has total mass 1, and all we have to show is that $\sigma = \mu_F$. We use (2.15) according to which if

$$U^\mu(x) = \int \log \frac{1}{|x-t|} d\mu(t)$$

is the logarithmic potential of the measure μ, then

$$U^{\mu_{F_k}}(x) = \log \frac{1}{\operatorname{cap}(F_k)} \tag{10.51}$$

for all $x \in F_k$ with the exception of a set of measure 0, and the same is true for E. Since $F \subset F_k$, it follows that

$$U^{\mu_F}(x) \le U^{\mu_{F_k}}(x) + \log \frac{\operatorname{cap}(F_k)}{\operatorname{cap}(F)}$$

for all $x \in F$ with the exception of a set of measure 0, and since every set of zero capacity has zero μ_F-measure (see [30, Remark I.1.7, p. 28]), it follows that this inequality is true μ_F-almost everywhere. But then by the principle of domination

[30, Theorem II.3.2] the same inequality is true for all $x \in \mathbf{C}$. Fixing such an $x \notin J$ and letting k tend to infinity through the subsequence $\{k_l\}$ we get from $\operatorname{cap}(F_k) \to \operatorname{cap}(F)$ that

$$U^{\mu_F}(x) \leq U^{\sigma}(x).$$

Thus, this inequality is true for all $x \in \mathbf{C} \setminus J$. However, the function

$$U^{\sigma}(x) - U^{\mu_F}(x)$$

vanishes at infinity, so it is harmonic there, and by an appeal to the minimum principle on the domain $\overline{\mathbf{C}} \setminus J$ yields that we must have

$$U^{\mu_F}(x) \equiv U^{\sigma}(x), \qquad x \in \mathbf{C} \setminus J.$$

Now we can conclude $\sigma = \mu_F$ from the unicity theorem [30, Theorem II.4.13].

We show that the claim in Lemma 10.8 follows from the just proven weak* convergence, and this proof incidentally also shows that equilibrium measures μ_{F_k} and μ_F are absolutely continuous in I with respect to linear Lebesgue measure, hence it is legitimate to talk about ω_{F_k} and ω_F there, and these are actually C^∞ functions (these same facts also follow from the discussion in Section 2.1 for the related results on balayage measures).

First of all let us notice that if $I \subset F$ is an interval, and $I^* \subset I$ is a subinterval such that even its closure lies in the interior of I, then μ_F is absolutely continuous in I, and its density is a C^∞ function on I^* with a bound on its derivative that depends only on I^* and I. In fact, we know that if $I = [a, b]$, then

$$d\mu_I(x) = \frac{1}{\pi\sqrt{(x-a)(b-x)}}dx,$$

and also that

$$d\mu_I(x) = d\mu_F\Big|_I(x) + \left(\frac{1}{\pi}\int_{\mathbf{R}\setminus I} \frac{|\sqrt{(u-a)(u-b)}|}{|x-u|\sqrt{(x-a)(b-x)}}d\mu_F(u)\right)dx,$$

where, in the last formula we used the fact that μ_I is the balayage of μ_F onto I, and we also used the formula (2.27) for this balayage measure. The preceding two formulae prove our claim.

Since $I \subset F_k \subset J$, we can use the same fact for all the measures μ_{F_k}. Thus, in I each of these measures is given by a density function ω_{F_k} with $|\omega'_{F_k}(x)| \leq C$, $x \in I^*$ for some constant C. The same is true of ω_F and so if it happens that $|\omega_{F_k}(x_k) - \omega_F(x_k)| \geq c > 0$, say $\omega_{F_k}(x_k) \geq \omega_F(x_k) + c$ for some $x_k \in I^*$ and for infinitely many k, say for $k \in \mathcal{N} \subset \mathbf{N}$, then we have the inequality

$$\omega_{F_k}(y) \geq \omega_F(y) + c/2 \qquad \text{for all} \quad |y - x_k| \leq c/2C, \quad y \in I^* \tag{10.52}$$

for all $k \in \mathcal{N}$. However, the weak* convergence of ω_{F_k} to ω_F implies that for each subinterval $L \subset I^*$ we must have

$$\int_L \omega_{F_k} \to \int_L \omega_F \qquad \text{as} \quad k \to \infty,$$

and so

$$\int_L \omega_{F_k} \leq \int_L \omega_F + c|L|/4 \tag{10.53}$$

is satisfied for sufficiently large k. Divide now I^* into finitely many such subintervals L, say $L_1, \ldots, L_s$ of length smaller than $c/4C$. Then (10.53) is true for all $L = L_j$, $j = 1, \ldots, s$ if k is sufficiently large. However, for every $k \in \mathcal{N}$ at least one of the intervals L_j, say L_{j_k} is a subset of the interval $\{x \mid |y - x_k| \le c/2C\}$, and so by integrating (10.52) on $L = L_{j_k}$ we get an inequality that contradicts (10.53). This contradiction proves that there cannot be infinitely many x_k's with $|\omega_{F_k}(x_k) - \omega_F(x_k)| \ge c > 0$, i.e. we must have $\omega_{F_k}(x) \to \omega_F(x)$ uniformly on I^*, and this was to be proven.

■

11 Appendix

11.1 Discretizing logarithmic integrals

In this appendix a discretization technique is used for constructing polynomials with prescribed properties. The technique goes back to [38].

Theorem 11.1 *Let $E \subseteq [0,1]$ be a compact set such that $E = \{0\} \bigcup \cup [a_i, b_i]$ with $0 < \cdots < a_2 < b_2 < a_1 < b_1 \leq 1$. Then for each j there is a constant A_j such that if $\tau \leq b_{j+2}$ and $E_{\tau,j} = [\tau, b_{j+1}] \bigcup \cup_{i=1}^{j} [a_i, b_i]$, then for all $n \geq 0$, n not necessarily an integer, there are polynomials P_n of degree at most n with all their zeros lying in $E_{\tau,j}$ such that*

$$P_n(0) \geq e^{n g_{\mathbf{C} \setminus E_{\tau,j}} \mu(0)} \tag{11.1}$$

and

$$|P_n(x)| \leq A_j, \qquad x \in E_{\tau,j}. \tag{11.2}$$

In particular, no matter how slowly the function $M(\tau)$ tends to infinity as $\tau \to 0$, there is a function $\tau \to \gamma_\tau$ tending to 0 as $\tau \to 0$ such that if $E_\tau = (E \cup [0, \gamma_t]) \setminus [0, \tau)$, then for all n, n not necessarily an integer, there are polynomials P_n of degree at most n with all their zeros lying in E_τ such that

$$P_n(0) \geq e^{n g_{\mathbf{C} \setminus E_\tau}(0)} \tag{11.3}$$

and

$$|P_n(x)| \leq M(\tau), \qquad x \in E_\tau. \tag{11.4}$$

For the proof we need the following lemma.

Lemma 11.2 *Let δ be a positive number. Then there is a constant C_δ depending only on δ for which the following is true. If $I = [a,b] \subset [0,1]$ is a subinterval of $[0,1]$, $|I| \geq \delta$, $d\mu(t) = v(t)dt$ is a measure on I with the properties*

$$v(t) \leq \frac{1}{\delta} \frac{1}{\sqrt{(t-a)(b-t)}}, \qquad t \in I \tag{11.5}$$

$$v(t) \geq \delta \frac{1}{\sqrt{(t-a)(b-t)}}, \qquad t \in I \tag{11.6}$$

and

$$\int_I v(t)dt =: \rho, \tag{11.7}$$

then for every $n \geq 0$ (n not necessarily an integer) there is a polynomial P_n of degree at most n such that

$$P_n(0) \geq e^{-nU^\mu(0)/\rho} \tag{11.8}$$

and

$$|P_n(x)| \leq C_\delta e^{-nU^\mu(x)/\rho}, \qquad x \in [0,1]. \tag{11.9}$$

Based on this lemma, the proof of Theorem 11.1 is straightforward. In fact, let $\mu_{E_{\tau,j}}$ be the equilibrium measure of $E_{\tau,j}$, and let $I_0 = [\tau, b_{j+1}]$, $I_i = [a_i, b_i]$, $1 \le i \le j$ be the subintervals of $E_{\tau,j}$. We shall apply the lemma with $I = I_i$, $\mu_i = \mu_{E_{\tau,j}}\big|_{I_i}$, $\rho_i = \mu_{E_{\tau,j}}(I_i)$ and with $n_i = n\rho_i$, $i = 0, \ldots, j$. Suppose we can do that, i.e. the conditions of the lemma are satisfied with some $\delta = \delta_i$ independent of $\tau \le b_{j+1}$. Thus, we obtain polynomials $P_{n\rho_i}$ of degree at most $n\rho_i$ such that

$$P_{n\rho_i}(0) \ge e^{-n\rho_i n U^{\mu_i}(0)/\rho_i} \tag{11.10}$$

$$|P_{n\rho_i}(x)| \le C_{\delta_i} e^{-n\rho_i U^{\mu_i}(x)/\rho_i}, \qquad x \in [0,1]. \tag{11.11}$$

Then for the product polynomial $P_n^*(x) = \prod_{i=0}^n P_{n\rho_i}(x)$ of degree at most $n\rho_0 + \cdots + n\rho_j = n$ we have

$$P_n^*(0) \ge \exp(-nU^{\mu_{E_{\tau,j}}}(0)) \tag{11.12}$$

and

$$|P_n^*(x)| \le A_j^* \exp(-nU^{\mu_{E_{\tau,j}}}(x)). \qquad x \in [0,1]. \tag{11.13}$$

Since (see (2.18))

$$g_{\mathbf{C}\setminus E_{\tau,j}}(x) = \log \frac{1}{\operatorname{cap}(E_{\tau,j})} - U^{\mu_{E_{\tau,j}}}(x),$$

and this is 0 for $x = E_{\tau,j}$, it follows that the polynomials

$$P_n(x) = \frac{1}{\operatorname{cap}(E_{\tau,j})} P_n(x)$$

satisfy both (11.1) and (11.2).

Thus, to complete the proof (pending the proof of the lemma) it is enough to show that the lemma is applicable for all $I = I_i$, $\mu_i = \mu_{E_{\tau,j}}\big|_{I_i}$, $\rho_i = \mu_{E_{\tau,j}}(I_i)$ and with $n_i = n\rho_i$, $i = 0, \ldots, j$, i.e. the conditions (11.5) and (11.6) of the lemma are satisfied with some $\delta = \delta_i$. But that is easy. In fact, if $i = 1, \ldots, j$, then for $t \in I_i$

$$d\mu_i(t) \le d\mu_{I_i}(t) = \frac{dt}{\pi\sqrt{(t-a_i)(b_i-t)}},$$

which is (11.5). The same works for $i = 0$.

On the other hand, for $i = 1, \ldots, j$ (c.f. the form (2.17) for the density of the equilibrium measure for sets consisting of several intervals)

$$d\mu_i(t) \ge d\mu_{[0,b_{i+1}]\cup[a_i,1]} \ge c_i \frac{dt}{\sqrt{t-a_i}},$$

and in a similar manner

$$d\mu_i(t) \ge d\mu_{[0,b_i]\cup[a_{i-1},1]} \ge c_i \frac{dt}{\sqrt{b_i - t}},$$

and these prove (11.6). The latter proof also works for $i = 0$;

$$d\mu_0(t) \ge d\mu_{[0,b_{j+1}]\cup[a_j,1]} \ge c_0 \frac{dt}{\sqrt{b_{j+1} - t}},$$

and in a similar manner for the left endpoint τ of I_0, we get

$$d\mu_0(t) \geq d\mu_{[\tau,1]} \geq c_i \frac{dt}{\sqrt{t-\tau}}.$$

With this we have shown that Lemma 11.2 can be applied, and the proof of the first part of Theorem 11.1 is complete pending the proof of Lemma 11.2.

The last statement in the theorem immediately follows from the first one if we set $\gamma_t = b_{j+1}$ for $\tau \in [\alpha_{j+1}, \alpha_j)$, where $\alpha_j < b_{j+2}$ goes to zero so fast that for $\tau < \alpha_j$ we have $A_j < M(\tau)$.

■

Proof of Lemma 11.2. First of all, since

$$\int_a^b \frac{1}{\sqrt{(t-a)(b-t)}} dt = \pi,$$

we have $\pi\delta \leq \rho \leq \pi/\delta$. Now by considering $v(t)/\rho$ instead of $v(t)$ for which (11.5) – (11.6) hold with δ replaced by δ^2/π, we may suppose

$$\int_I v(t)dt = \rho = 1,$$

i.e. that μ has total mass 1.

The proof of the lemma is carried out by a discretization of the measure μ. This proof is an adaptation of the proof of [30, Theorem VI.4.2].

First let $n \geq 2$ be an *integer* (for $n = 0$ set $P_0 = 1$ while for $n = 1$ set $P_1(x) = x - \xi$, where ξ is the weight point of μ, see below). We have to construct polynomials

$$P_n(x) = \prod_{j=0}^{n-1} (x - \xi_j)$$

such that

$$-\log|P_n(0)| - nU^\mu(0) \leq 0 \tag{11.14}$$

and

$$-\log|P_n(x)| - nU^\mu(x) \geq C_\delta \tag{11.15}$$

for all $x \in [0,1]$, where C_δ depends only on δ. Partition $I =: [a,b]$ by the points $a = t_0 < t_1 < \ldots < t_n = b$ into n intervals $I_j = [t_j, t_{j+1}]$, $j = 0, 1, \ldots, n-1$ with $\mu(I_j) = 1/n$, and let ξ_j be the weight point of the restriction of μ to I_j; i.e.

$$\xi_j = n \int_{I_j} tv(t)\, dt. \tag{11.16}$$

Actually, ξ_i depends on i i.e. $\xi_i = \xi_{i,n}$, but for simpler notation we just write $\xi_i = \xi_{i,n}$. Set

$$P_n(t) = \prod_{j=0}^{n-1} (t - \xi_j).$$

We claim that these P_n satisfy (11.14) – (11.15).

We write

$$-\log|P_n(x)| - nU^\mu(x) = \sum_{j=0}^{n-1} n\int_{I_j} \log\left|\frac{x-t}{x-\xi_j}\right| v(t)dt =: \sum_{j=0}^{n-1} L_j(x). \tag{11.17}$$

The proof of (11.14) is immediate: for $x \notin \mathrm{Int}(I)$, in particular for $x = 0$, the function $t \to \log|x-t|$ is concave on I_j, so

$$n\int_{I_j} \log|x-t|v(t)dt \le \log|x-\xi_j|,$$

and hence every term $L_j(x)$ in (11.17) is at most 0, and this proves (11.14).

Before we can prove (11.15), we have to make several estimations for the numbers t_j. First of all for $j = 1, \dots, n$ if $t_j \in [a, (a+b)/2]$, then we have

$$\frac{j}{n} = \int_a^{t_j} v(t)dt \le \frac{1}{\delta}\int_a^{t_j} \frac{1}{\sqrt{t-a}\sqrt{\delta/2}}dt \le \frac{2\sqrt{2}}{\delta^{3/2}}\sqrt{t_j - a},$$

and the same estimate is automatically satisfied if $t_j \in [(a+b)/2, b]$. From the other side

$$\frac{j}{n} = \int_a^{t_j} v(t)dt \ge \delta\int_a^{t_j} \frac{1}{\sqrt{t-a}}dt \ge 2\delta\sqrt{t_j - a}.$$

These give

$$\frac{\delta^3}{8}\left(\frac{j}{n}\right)^2 \le t_j - a \le \frac{1}{4\delta^2}\left(\frac{j}{n}\right)^2, \tag{11.18}$$

for $j = 1, \dots, n$, and by symmetry

$$\frac{\delta^3}{8}\left(\frac{n-j}{n}\right)^2 \le b - t_j \le \frac{1}{4\delta^2}\left(\frac{n-j}{n}\right)^2 \tag{11.19}$$

for all $j = 0, \dots, n-1$. Thus, on $I_j = [t_j, t_{j+1}]$, $j = 1, \dots, n-2$ we have

$$\delta \cdot 2\delta\frac{n}{j+1}2\delta\frac{n}{n-j} \le v(t) \le \frac{1}{\delta}\frac{2\sqrt{2}}{\delta^{3/2}}\frac{n}{j}\frac{2\sqrt{2}}{\delta^{3/2}}\frac{n}{n-1-j}. \tag{11.20}$$

Combining these inequalities with $\int_{I_j} v = 1/n$ we get for $j = 1, \dots, n-2$

$$\frac{\delta^4}{8}\frac{j(n-1-j)}{n^3} \le |I_j| \le \frac{1}{4\delta^3}\frac{(j+1)(n-j)}{n^3} \tag{11.21}$$

while for $j = 0$ and $j = n-1$ we can see from the estimates on t_1 and t_{n-1} that

$$\frac{\delta^3}{8}\left(\frac{1}{n}\right)^2 \le |I_0|, |I_{n-1}| \le \frac{1}{4\delta^2}\left(\frac{1}{n}\right)^2. \tag{11.22}$$

In particular, for all j we have

$$\frac{\delta^7}{8} \le \frac{|I_{j+1}|}{|I_j|} \le \frac{8}{\delta^7}. \tag{11.23}$$

These give for $k < l \le n/2$

$$\begin{aligned}|I_{k+1}| + |I_{k+2}| + \cdots + |I_l| &\ge \frac{\delta^4}{8}\sum_{j=k+1}^{l}\frac{j(n-1-j)}{n^3} \ge \frac{\delta^4}{16n^2}\sum_{j=k+1}^{l} j \\ &= \frac{\delta^4}{32n^2}(l(l+1)-k(k+1)) \ge \frac{\delta^4}{32}\frac{(l-k)l}{n^2}.\end{aligned} \tag{11.24}$$

In a similar fashion, for $n/2 \le k < l$ we have

$$|I_{k+1}| + |I_{k+2}| + \cdot + |I_l| \ge \frac{\delta^4}{32}\frac{(l-k)(n-k)}{n^2}. \tag{11.25}$$

Finally, for $k \le n/2 \le l$ by splitting the sum from k to $[n/2]$ and from $[n/2]+1$ to l we the preceding inequalities easily yield

$$|I_{k+1}| + |I_{k+2}| + \cdot + |I_l| \ge \frac{\delta^4}{64}\frac{(l-k)n}{n^2}. \tag{11.26}$$

After these preparations let us return to the proof of (11.15). It is enough to prove it only for $x \in I$. In fact, the left hand side is harmonic in $\mathbf{C} \setminus I$ including the point infinity, therefore (11.15) for $x \in I$ implies its validity for all $x \in \mathbf{C}$. Thus, let $x \in I_{j_0}$ for some $j_0 = 0, 1, \ldots, n-1$.

First we prove that the individual terms in (11.17) are (uniformly) bounded from below by a constant depending only on δ. In fact, this is clear for $j \ne j_0, j_0 \pm 1$, for then by (11.23) for $t, \xi_j \in I_j$

$$\left|\frac{x-t}{x-\xi_j}\right| \ge \frac{\delta^7}{16}.$$

On the other hand, for $j = j_0, j_0 \pm 1$ we have

$$\left|\frac{x-t}{x-\xi_j}\right| \ge \frac{\delta^7}{16}\left|\frac{x-t}{|I_j|}\right|,$$

and hence

$$L_j(x) = n\int_{I_j}\log\left|\frac{x-t}{x-\xi_j}\right| v(t)dt \ge \log\frac{\delta^7}{16} + n\int_{I_j}\log\left|\frac{x-t}{|I_j|}\right| v(t)dt.$$

Let first $j \ne 0$ or $j \ne n-1$. Apply on the interval I_j the upper estimate for $v(t)$ from (11.20), make the substitution $t = t_j + |I_j|u$, $x = t_j + |I_j|y$, and for the factor $|I_j|$ in $dt = |I_j|du$ apply (11.21). What we obtain is

$$L_j(x) \ge \log\frac{\delta^7}{16} + \int_0^1 \log|y-u|du\frac{8}{\delta^7}.$$

When $j = 0$ or $j = n$, the argument is similar, only we arrive at a multiple of the integral

$$\int_0^1 \frac{1}{\sqrt{u}}\log|y-u|du.$$

In either case the uniform lower boundedness of the terms $L_j(x)$ follow.

It follows from (11.21) – (11.22) and (11.24) – (11.26) that there is an $L \geq 1$ depending only on δ such that if $x \in I_{j_0}$ and $t \in I_j$ with $|j - j_0| \geq L$, then we have

$$\frac{\xi_j - t}{x - \xi_j} \geq -\frac{1}{2}. \tag{11.27}$$

From previous discussion on the lower boundedness of individual terms, for $|j-j_0| < L$ we can write $L_j(x) \geq -D_1$ for a D_1 that depends only on δ. Hence

$$\sum_{|j-j_0|<L} L_j(x) \geq -2D_1 L. \tag{11.28}$$

For other j's, (11.27) holds, and we can write, for $x \in I_{j_0}$ and $|j - j_0| \geq L$, the integrand in $L_j(x)$ (see (11.17)) as

$$\log\left|1 + \frac{\xi_j - t}{x - \xi_j}\right| = \frac{\xi_j - t}{x - \xi_j} + O\left(\left|\frac{\xi_j - t}{x - \xi_j}\right|^2\right).$$

Thus, we have for such j's

$$\begin{aligned} L_j(x) &= n\int_{I_j} O\left(\left|\frac{\xi_j - t}{x - \xi_j}\right|^2\right) v(t)dt \\ &= O\left(\frac{|I_j|^2}{(\xi_j - \xi_{j_0})^2}\right), \end{aligned} \tag{11.29}$$

because the integrals

$$\int_{I_j} \frac{\xi_j - t}{x - \xi_j} v(t)dt$$

vanish by the choice of the points ξ_j. Here the constant in O is an absolute constant.

In view of (11.28) we have to bound the sums

$$S_1(x) = \sum_{j=1}^{j_0-L} |L_j(x)|$$

and

$$S_2(x) = \sum_{j=j_0+L}^{n-1} |L_j(x)|,$$

and since these are similar sums, let us deal only with the first one. Since for $j \leq j_0 - L$ we have

$$|\xi_j - \xi_{j_0}| \geq |I_{j+1}| + |I_{j+2}| + \cdot + |I_{j_0-1}|,$$

for $j_0 \leq n/2$ we get from (11.21) – (11.22) and (11.24)

$$L_j(x) = O\left(\frac{|I_j|^2}{(\xi_j - \xi_{j_0})^2}\right) = O\left(\left(\frac{(j+1)}{(j_0 - j)j_0}\right)^2\right) = O\left(\frac{1}{(j_0 - j)^2}\right).$$

If $j_0 > n/2 \geq j$, then instead of (11.24) apply (11.26) to conclude

$$L_j(x) = O\left(\frac{|I_j|^2}{(\xi_j - \xi_{j_0})^2}\right) = O\left(\left(\frac{(j+1)}{(j_0 - j)n}\right)^2\right) = O\left(\frac{1}{(j_0 - j)^2}\right).$$

Finally, if $n/2 \le j \le j_0 - L$ then (11.21) – (11.22) and (11.25) yield

$$L_j(x) = O\left(\frac{|I_j|^2}{(\xi_j - \xi_{j_0})^2}\right) = O\left(\left(\frac{(n-j)}{(j_0-j)(n-j)}\right)^2\right) = O\left(\frac{1}{(j_0-j)^2}\right).$$

These show that S_1 is bounded by a constant multiple of $\sum 1/k^2$, therefore it is bounded.

This completes the proof, provided n is an integer.

If n is not an integer, then apply the just proven integer case with $[n]$, the integral part of n, and set $P_n(x) = P_{[n]}(x)$. Then

$$P_n(0) \ge e^{-[n]U^\mu(0)} \ge e^{-nU^\mu(0)},$$

because $U^\mu(0) \ge 0$, i.e. (11.8) is automatically satisfied. As for (11.9), notice that $\mu \le \frac{\pi}{\delta}\mu_I$ where μ_I denotes the equilibrium measure of I, and hence for $x \in [0,1]$

$$U^\mu(x) \le \frac{\pi}{\delta}U^{\mu_I}(x) \le \frac{\pi}{\delta}\log\frac{4}{|I|} \le \frac{\pi}{\delta}\log\frac{4}{\delta},$$

and so if C'_δ is the constant in (11.9) for integer values of n, then

$$|P_n(x)| \le C'_\delta e^{-[n]U^\mu(x)} \le C'_\delta e^{-nU^\mu(x)}e^{U^\mu(x)} \le C'_\delta\left(\frac{4}{\delta}\right)^{4/\delta}e^{-nU^\mu(x)}.$$

This proves (11.9), and the proof of the lemma is over.

■

11.2 Some lemmas on Θ integrals

This appendix contains some elementary results on the integrals involving Θ_E or Θ_E^* used in this work. Thus, let E be compact, and $\Theta = \Theta_E$ or $\Theta = \Theta_E^*$, see (2.1) and (4.1). We shall work with Θ_E, the case Θ_{E^*} is perfectly analogous. We may assume E to lie in the unit disk.

We shall frequently use that if $A > 1$ then for any $0 < a < 1$

$$\begin{aligned} c_A \sum_{A^{-k}\ge a} \left(A^k\Theta(A^{-k})\right)^2 + O_A(1) &\le \int_a^1 \frac{\Theta(t)^2}{t^3}dt \\ &\le C_A \sum_{A^{-k}\ge a} \left(A^k\Theta(A^{-k})\right)^2 + O_A(1), \end{aligned} \tag{11.30}$$

where (and in what follows) c_A, C_A are positive constants depending only on A, and $O_A(1)$ denotes a term that is less than a constant depending only on A. Let us also agree that sums like $\sum_{A^{-k}\ge a}$ are taken for those k for which $a \le A^{-k} \le 1$. The preceding inequality is standard and immediately follows, since Θ is an increasing function, and so

$$\begin{aligned} (A^k)^3(A-1)A^{-k-1}\Theta(A^{-k-1})^2 &\le \int_{A^{-k-1}}^{A^{-k}} \frac{\Theta(t)^2}{t^3}dt \\ &\le (A^{k+1})^3(A-1)A^{-k-1}\Theta(A^{-k})^2, \end{aligned}$$

i.e.

$$\frac{3(A-1)}{A^2}(A^{k+1}\Theta(A^{-k-1}))^2 \le \int_{A^{-k-1}}^{A^{-k}} \frac{\Theta(t)^2}{t^3}dt \le A^2 3(A-1)(A^{-k}\Theta(A^{-k})^2.$$

Now take the sum for $A^{-k} \ge a$. Note also that since $\Theta(t) \le t$, the individual terms $(A^{-k}\Theta(A^{-k}))^2$ are bounded by 1.

Lemma 11.3 *For $A > 1$ set*

$$\overline{\Theta} = \overline{\Theta}_E^{\{A\}}(t) = \Theta_E(t) - \Theta_E(t/A).$$

Then for $0 < a < 1$

$$\begin{aligned} c_A \sum_{A^{-k}\ge a} \left(A^k\overline{\Theta}(A^{-k})\right)^2 + O_A(1) &\le \int_a^1 \frac{\Theta(t)^2}{t^3}dt \\ &\le C_A \sum_{A^{-k}\ge a} \left(A^k\overline{\Theta}(A^{-k})\right)^2 + O_A(1). \end{aligned} \tag{11.31}$$

Proof. Using that

$$\Theta(A^{-k}) = \sum_{l=k}^{\infty} \overline{\Theta}(A^{-l}),$$

we can derive from Jensen's inequality

$$\begin{aligned} \int_a^1 \frac{\Theta(t)^2}{t^3}dt &\le C_A \sum_{A^{-k}\ge a} \left(A^k\Theta(A^{-k})\right)^2 + O_A(1) \\ &= C_A \sum_{A^{-k}\ge a} \left(\sum_{l=k}^{\infty} \frac{1}{A^{2(l-k)}}(A^l\overline{\Theta}(A^{-l})^2\right)^2 + O_A(1) \\ &\le C_A \sum_{A^{-k}\ge a} \sum_{l=k}^{\infty} \frac{1}{A^{2(l-k)}}(A^l\overline{\Theta}(A^{-l})^2 + O_A(1) \le \end{aligned}$$

$$C_A \sum_{A^{-l}\ge a} (A^l\overline{\Theta}(A^{-l})^2 + C_A \sum_{A^{-l}<a} (A^l\overline{\Theta}(A^{-l})^2 \left(\sum_{A^{-k}\ge a} \frac{1}{A^{2(l-k)}}\right) + O_A(1).$$

Here the very last sum in the brackets is at most C_A/a^2A^{2l}, therefore the second term on the right is at most

$$\frac{C_A}{a^2} \sum_{A^{-l}<a} \overline{\Theta}(A^{-l})^2 \le \frac{C_A}{a^2} \sum_{A^{-l}<a} A^{-2l} \le C_A,$$

where we used $\overline{\Theta}(A^{-l}) \le A^{-l}$. This proves the right inequality in (11.31).

The proof of the left inequality is simpler:

$$\sum_{A^{-k}\ge a} \left(A^k\overline{\Theta}(A^{-k})\right)^2 \le \sum_{A^{-k}\ge a} \left(A^k\Theta(A^{-k})\right)^2 \le C_A \int_a^1 \frac{\Theta(t)^2}{t^3}dt + O_A(1),$$

where in the last step we used (11.30).

■

Lemma 11.4 *Let $A > 1$ and $\overline{\Theta}(t) = \Theta(t) - \Theta(t/A)$. Then for $0 < a < 1$*

$$\begin{aligned} c_A \sum_{A^{-k} \geq a} \left(A^k \overline{\Theta}(A^{-k})\right)^2 + O_A(1) &\leq \int_a^1 \frac{\overline{\Theta}(t)^2}{t^3} dt \\ &\leq C_A \sum_{A^{-k} \geq a} \left(A^k \overline{\Theta}(A^{-k})\right)^2 + O_A(1). \end{aligned} \tag{11.32}$$

Since $\overline{\Theta}$ may not be monotone, this does not follow immediately from (11.30) and (11.31).

As a corollary of 11.32 and (11.31) we get

$$\int_a^1 \frac{\overline{\Theta}(t)^2}{t^3} dt \leq \int_a^1 \frac{\Theta(t)^2}{t^3} dt \leq C \int_a^1 \frac{\overline{\Theta}(t)^2}{t^3} dt + O(1), \tag{11.33}$$

Proof. We have

$$\overline{\Theta}(t) = |[t/A, t] \setminus E|,$$

therefore if $t \in [A^{-k-1}, A^{-k}]$, then

$$\overline{\Theta}(A^{-k}) \leq \overline{\Theta}(t) + \overline{\Theta}((A-1)A^{-k-1}) + t).$$

This yields

$$\begin{aligned} (A^k \overline{\Theta}(A^{-k}))^2 &\leq A^{2k} \frac{A^{k+1}}{A-1} \int_{A^{-k-1}}^{(2A-1)A^{-k-1}} 2(\overline{\Theta}(t))^2 dt \\ &\leq C_A \int_{A^{-k-1}}^{(2A-1)A^{-k-1}} \frac{\overline{\Theta}(t)^2}{t^3} dt. \end{aligned}$$

On taking sum for $A^{-k} \geq a$ we obtain the left inequality in (11.32).

The proof of the right inequality is similar if we use that for $t \in [A^{-k-1}, A^{-k}]$ we have

$$\overline{\Theta}(t) \leq \overline{\Theta}(A^{-k-1}) + \overline{\Theta}(A^{-k}).$$

■

Lemma 11.5 *Suppose $E \subset [0,1]$ is compact, and with some $s > 0$ set*

$$\tilde{E} = \{t^s \,|\, t \in E.\}$$

Then for $0 < a < 1$ we have

$$c \int_a^1 \frac{\Theta_E(t)^2}{t^3} dt + O(1) \leq \int_{a^s}^1 \frac{\Theta_{\tilde{E}}(t)^2}{t^3} dt \leq C \int_a^1 \frac{\Theta(t)^2}{t^3} dt + O(1). \tag{11.34}$$

Proof. Using Lemma 11.3 with $A = 2^s$ we can write

$$\int_{a^s}^{1} \frac{\Theta_{\tilde{E}}(t)^2}{t^3} dt \leq C_s \sum_{2^{-ks} \geq a^s} \left(2^{ks} \overline{\Theta}_{\tilde{E}}^{\{2^s\}}(2^{-ks})\right)^2 + O_s(1).$$

Here for $s \geq 1$

$$2^{-(k+1)(s-1)} \overline{\Theta}_{E}^{\{2\}}(2^{-k}) \leq \overline{\Theta}_{\tilde{E}}^{\{2^s\}}(2^{-ks}) \leq 2^{-k(s-1)} \overline{\Theta}_{E}^{\{2\}}(2^{-k}) \tag{11.35}$$

because on the interval $[2^{-k-1}, 2^{-k}]$ the mapping $t \to t^s$ is a contraction with contraction factor $\geq 2^{-(k+1)(s-1)}$ and $\leq 2^{-k(s-1)}$.

Thus,

$$\int_{a^s}^{1} \frac{\Theta_{\tilde{E}}(t)^2}{t^3} dt \leq C_s \sum_{2^{-k} \geq a} \left(2^{k} \overline{\Theta}_{E}^{\{2\}}(2^{-k})\right)^2 + O_s(1),$$

and an application of Lemma 11.3 gives the right inequality in (11.32).

The proof of the left inequality is perfectly similar if we use the lower bound in (11.35).

Finally, if $0 < s < 1$, then the only modification is that instead of (11.35) we have

$$2^{-k(s-1)} \overline{\Theta}_{E}^{\{2\}}(2^{-k}) \leq \overline{\Theta}_{\tilde{E}}^{\{2^s\}}(2^{-ks}) \leq 2^{-(k+1)(s-1)} \overline{\Theta}_{E}^{\{2\}}(2^{-k}).$$

■

12 References

[1] L. V. Ahlfors, *Conformal Invariants*, McGraw-Hill Series in Higher Matheamtics, McGraw-Hill, New York 1973

[2] V. V. Andrievskii, A Remez-type inequality in terms of capacity, *Complex Variables Theory Appl.*, **45**(2001), 35–46.

[3] V. V. Andrievskii, The Nikol'skii-Timan-Dzjadyk theorem for functions on compact sets of the real line, *Constr. Approx.*, **17**(2001), 431–454.

[4] V. V. Andrievskii, *Application of conformal and quasiconformal mappings and their properties in approximation theory*, Handbook of complex analysis: geometric function theory, Vol. 1, 493–520, North-Holland, Amsterdam, 2002.

[5] V. V. Andrievskii, Uniformly Perfect Subsets of the Real Line and John Domains, *Comput. Methods Funct. Theory*, **3**(2003), 385-396.

[6] V. V. Andrievskii, The highest smoothness of the Green function implies the highest density of a set, *Arkiv för Matematik*, **42**(2004), 217-238.

[7] V. V. Andrievskii, On the Green function for a complement of a finite number of real intervals, *Constr. Approx.*, **20**(2004), 4, 565–583.

[8] V. V. Andrievskii V.V., On sparse sets with the Green function of the highest smoothness, *Comput. Methods Funct. Theory* (to appear)

[9] V. V. Andrievskii, On optimal smoothness of the Green function for the complement of a Cantor-type set, *Constr. Approx.* (to appear)

[10] A. Baernstein II, Integral means, univalent function and cicular symmetrization, *Acta Math.*, **133**(1975), 139–169.

[11] S. N. Bernstein, Sur la meilleure approximation de $|x|$ par des polynomes des degrés donnés, *Acta Math.*, **37**(1914), 1–57.

[12] S. N. Bernstein, On the best approximation of $|x|^p$ by means of polynomials of extremely high degree, *Izv. Akad. Nauk SSSR,* Ser. Mat. **2**(1938), 160– 180. Reprinted in S. N. Bernstein *Collected Works*, Vol. 2, pp. 262–272. Izdat. Nauk SSSR, Moscow, 1954 (Russian).

[13] S. N. Bernstein, On the best approximation of $|x-c|^p$, *Dokl. Akad. Nauk SSSR,* **18**(1938), 379–384. Reprinted in S. N. Bernstein *Collected Works*, Vol. 2, pp. 273–260. Izdat. Nauk SSSR, Moscow, 1954 (Russian).

[14] L. Białas and A. Volberg, Markov's propety of the Cantor ternary set, *Studia Math.,* **104**(1993), 259–268.

[15] L. Carleson and V. Totik, Hölder continuity of Green's functions, *Acta Sci. Math.*, **70**(2004), 557–608.

[16] A. B. Bogatyrev, Effective computation of Chebyshev polynomials for several intervals, *Math. USSR Sb.*, **190**(1999), 1571–1605.

[17] R. A. DeVore and G. G. Lorentz, *Contructive Approximation*, Grundlehren der mathematisschen Wissenschaften, **303**, Springer-Verlag, New York, 1993

[18] T. Erdélyi, A. Kroó and J. Szabados, Markov–Bernstein type inequalities on compact subsets of **R**, *Analysis Math.*, **26**(2000), 17–34.

[19] J. Geronimo and W. Van Assche, Orthogonal polynomials on several intervals via a polynomial mapping, *Trans. Amer. Math. Soc.*, **308**(1988), 559–581.

[20] K. G. Ivanov and V. Totik, Fast decreasing polynomials, *Constr. Approx.*, **6**(1990), 1–20.

[21] N. S. Landkof: *Foundations of Modern Potential Theory*, Grundlehren der Mathematischen Wissenschaften, **190**, Springer–Verlag, New York, 1972

[22] V. G. Maz'ja, Regularity at the boundary of solutions of elliptic equations, and conformal mappings, *Dokl. Akad. Nauk SSSR*, **152**(1963), 1297–1300. (Russian)

[23] V. G. Maz'ja, On the modulus of continuity of the solutions of the Dirichlet problem near irregular boundary, in: Problems in mathematical analysis; Boundary problems and integral equations, *Izdat. Leningrad. Univ., Leningrad*, 1966, 45–58. (Russian)

[24] V. G. Maz'ja, On the regularity of boundary points for elliptic equations, in: Investigations on linear operators and function theory; 99 unsolved problems in linear and complex analysis, *Zap. Nauchn. Sem. Leningrad. Otdel. Mat. Inst. Steklov. (LOMI)*, **81**(1978), 197–199. (Russian)

[25] V. G. Maz'ja, On the modulus of continuity of a harmonic function at a boundary point, *Zap. Nauchn. Sem. Leningrad. Otdel. Mat. Inst. Steklov. (LOMI)*, **135**(1984), 87–95. (Russian)

[26] P. Nevai and V. Totik, Weighted polynomial inequalities, *Constr. Approx.*, **2**(1986), 113–127.

[27] R. Nevanlinna, *Analytic Functions*, Grundlehren der mathematischen Wissenschaften, **162**, Springer Verlag, Berlin, 1970

[28] F. Peherstorfer, Deformation of minimizing polynomials and approximation of several intervals by an inverse polynomial mapping, *J. Approx. Theory*, **111**(2001), 180–195.

[29] T. Ransford, *Potential Theory in the Complex Plane*, Cambridge University Press, Cambridge, 1995

[30] E. B. Saff and V. Totik, *Logarithmic Potentials with External Fields*, Grundlehren der mathematischen Wissenschaften, **316**, Springer Verlag, Berlin, Heidelberg, 1997.

[31] H. Stahl and V. Totik, *General Orthogonal Polynomials* Encyclopedia of Mathematics, **43**, Cambridge University Press, New York 1992

[32] P. M. Tamrazov, *Smoothness and Polynomial Approximation*, Naukova Dumka, Moscow, 1975, 271 pp. (Russian)

[33] F. Toókos, Smoothness of Green's functions and density of sets, *Acta Sci. Math* (Szeged), **71**(2005), 117–146.

[34] F. Toókos, A Wiener-type condition for Hölder continuity, *Acta Math. Hungarica*, (to appear)

[35] M. Tsuji, *Potential Theory in Modern Function Theory*, Maruzen, Tokyo, 1959

[36] V. Totik, Fast decreasing polynomials via potentials, *J. D'Analyse Math.*, **62**(1994), 131–154.

[37] V. Totik, Markoff constants for Cantor sets, *Acta Sci. Math. (Szeged)*, **60**(1995), 715–734.

[38] V. Totik, Polynomial inverse images of intervals and polynomial inequalities, *Acta Math.*, **187**(2001), 139–160.

[39] V. Totik, On Markoff's inequality, *Constr. Approx.*, **18**(2002), 427–441.

[40] R. Varga and A. J. Carpenter, On the Bernstein conjecture in approximation theory, *Constr. Approx.*, **1**(1985), 333-348.

[41] R. K. Vasiliev, Chebyshev Polynomials and Approximation Theory on Compact Subsetc of the Real Axis, Saratov University Publishing House, 1998.

[42] J. L. Walsh: *Interpolation and Approximation by Rational Functions In the Complex Domain*, Colloquium Publications **20**, Amer. Math. Soc., Providence 1960.

[43] H. Widom: Extremal polynomials associated with a system of curves in the complex plane, *Advances in Math.*, **2**(1969), 127–232.

13 List of symbols

$\mathrm{cap}(E)$ logarithmic capacity of the set E 6

$C_r(z)$ circle of radius r about z 4

$\mathcal{C} = \mathcal{C}(\varepsilon_1, \varepsilon_2, \ldots)$ Cantor set with parameters ε_i 44

δ_a Dirac delta 8

$\Delta_r(z)$ disk of radius r about z 4

E^c circular projection 0, 5

E^{symm} symmetrization of the set E 37

$\mathcal{E}_n(f, E)$ best approximation of f on E by polynomials of degree $\leq n$ 126

$g_{\mathbf{C}\setminus E}(z)$ Green function of $\mathbf{C} \setminus E$ 7

μ_E equilibrium measure of the set E 6

$\overline{\nu}$, $\widehat{\nu}$ balayage measures, 8

$\omega(z, F, G)$ harmonic measure 9

$\omega_E(x)$ density of the equilibrium measure 7

t_n length of the intervals at the n-th Cantor level 46

$\Theta_{E,P}(t)$ density function 4

U^ν logarithmic potential of the measure ν 6

$V(E)$ logarithmic energy of the set E 6

14 List of figures

15 Index

Editorial Information

To be published in the *Memoirs*, a paper must be correct, new, nontrivial, and significant. Further, it must be well written and of interest to a substantial number of mathematicians. Piecemeal results, such as an inconclusive step toward an unproved major theorem or a minor variation on a known result, are in general not acceptable for publication. Papers appearing in *Memoirs* are generally at least 80 and not more than 200 published pages in length. Papers less than 80 or more than 200 published pages require the approval of the Managing Editor of the Transactions/Memoirs Editorial Board.

As of July 31, 2006, the backlog for this journal was approximately 11 volumes. This estimate is the result of dividing the number of manuscripts for this journal in the Providence office that have not yet gone to the printer on the above date by the average number of monographs per volume over the previous twelve months, reduced by the number of volumes published in four months (the time necessary for preparing a volume for the printer). (There are 6 volumes per year, each containing at least 4 numbers.)

A Consent to Publish and Copyright Agreement is required before a paper will be published in the *Memoirs*. After a paper is accepted for publication, the Providence office will send a Consent to Publish and Copyright Agreement to all authors of the paper. By submitting a paper to the *Memoirs*, authors certify that the results have not been submitted to nor are they under consideration for publication by another journal, conference proceedings, or similar publication.

Information for Authors

Memoirs are printed from camera copy fully prepared by the author. This means that the finished book will look exactly like the copy submitted.

The paper must contain a *descriptive title* and an *abstract* that summarizes the article in language suitable for workers in the general field (algebra, analysis, etc.). The *descriptive title* should be short, but informative; useless or vague phrases such as "some remarks about" or "concerning" should be avoided. The *abstract* should be at least one complete sentence, and at most 300 words. Included with the footnotes to the paper should be the 2000 *Mathematics Subject Classification* representing the primary and secondary subjects of the article. The classifications are accessible from `www.ams.org/msc/`. The list of classifications is also available in print starting with the 1999 annual index of *Mathematical Reviews*. The Mathematics Subject Classification footnote may be followed by a list of *key words and phrases* describing the subject matter of the article and taken from it. Journal abbreviations used in bibliographies are listed in the latest *Mathematical Reviews* annual index. The series abbreviations are also accessible from `www.ams.org/publications/`. To help in preparing and verifying references, the AMS offers MR Lookup, a Reference Tool for Linking, at `www.ams.org/mrlookup/`. When the manuscript is submitted, authors should supply the editor with electronic addresses if available. These will be printed after the postal address at the end of the article.

Electronically prepared manuscripts. The AMS encourages electronically prepared manuscripts, with a strong preference for AMS-LaTeX. To this end, the Society has prepared AMS-LaTeX author packages for each AMS publication. Author packages include instructions for preparing electronic manuscripts, the *AMS Author Handbook*, samples, and a style file that generates the particular design specifications of that publication series. Though AMS-LaTeX is the highly preferred format of TeX, author packages are also available in AMS-TeX.

Authors may retrieve an author package from e-MATH starting from `www.ams.org/tex/` or via FTP to `ftp.ams.org` (login as `anonymous`, enter username as password, and type `cd pub/author-info`). The *AMS Author Handbook* and the *Instruction Manual* are available in PDF format following the author packages link from `www.ams.org/tex/`. The author package can also be obtained free of charge by sending email to tech-support@ams.org (Internet) or from the Publication Division, American Mathematical Society, 201 Charles St., Providence, RI 02904-2294, USA. When requesting an author package, please specify $\mathcal{A}_{\mathcal{M}}\mathcal{S}$-LaTeX or $\mathcal{A}_{\mathcal{M}}\mathcal{S}$-TeX and the publication in which your paper will appear. Please be sure to include your complete mailing address.

Sending electronic files. After acceptance, the source file(s) should be sent to the Providence office (this includes any TeX source file, any graphics files, and the DVI or PostScript file).

Before sending the source file, be sure you have proofread your paper carefully. The files you send must be the EXACT files used to generate the proof copy that was accepted for publication. For all publications, authors are required to send a printed copy of their paper, which exactly matches the copy approved for publication, along with any graphics that will appear in the paper.

TeX files may be submitted by email, FTP, or on diskette. The DVI file(s) and PostScript files should be submitted only by FTP or on diskette unless they are encoded properly to submit through email. (DVI files are binary and PostScript files tend to be very large.)

Electronically prepared manuscripts can be sent via email to pub-submit@ams.org (Internet). The subject line of the message should include the publication code to identify it as a Memoir. TeX source files, DVI files, and PostScript files can be transferred over the Internet by FTP to the Internet node e-math.ams.org (130.44.1.100).

Electronic graphics. Comprehensive instructions on preparing graphics are available at www.ams.org/jourhtml/graphics.html. A few of the major requirements are given here.

Submit files for graphics as EPS (Encapsulated PostScript) files. This includes graphics originated via a graphics application as well as scanned photographs or other computer-generated images. If this is not possible, TIFF files are acceptable as long as they can be opened in Adobe Photoshop or Illustrator. No matter what method was used to produce the graphic, it is necessary to provide a paper copy to the AMS.

Authors using graphics packages for the creation of electronic art should also avoid the use of any lines thinner than 0.5 points in width. Many graphics packages allow the user to specify a "hairline" for a very thin line. Hairlines often look acceptable when proofed on a typical laser printer. However, when produced on a high-resolution laser imagesetter, hairlines become nearly invisible and will be lost entirely in the final printing process.

Screens should be set to values between 15% and 85%. Screens which fall outside of this range are too light or too dark to print correctly. Variations of screens within a graphic should be no less than 10%.

Inquiries. Any inquiries concerning a paper that has been accepted for publication should be sent directly to the Electronic Prepress Department, American Mathematical Society, 201 Charles St., Providence, RI 02904, USA.

Editors

This journal is designed particularly for long research papers, normally at least 80 pages in length, and groups of cognate papers in pure and applied mathematics. Papers intended for publication in the *Memoirs* should be addressed to one of the following editors. In principle the Memoirs welcomes electronic submissions, and some of the editors, those whose names appear below with an asterisk (*), have indicated that they prefer them. However, editors reserve the right to request hard copies after papers have been submitted electronically. Authors are advised to make preliminary email inquiries to editors about whether they are likely to be able to handle submissions in a particular electronic form.

***Algebra** to ALEXANDER KLESHCHEV, Department of Mathematics, University of Oregon, Eugene, OR 97403-1222; email: ams@noether.uoregon.edu

Algebra and its application to MINA TEICHER, Emmy Noether Research Institute for Mathematics, Bar-Ilan University, Ramat-Gan 52900, Israel; email: teicher@macs.biu.ac.il

Algebraic geometry to DAN ABRAMOVICH, Department of Mathematics, Brown University, Box 1917, Providence, RI 02912; email: amsedit@math.brown.edu

***Algebraic number theory** to V. KUMAR MURTY, Department of Mathematics, University of Toronto, 100 St. George Street, Toronto, ON M5S 1A1, Canada; email: murty@math.toronto.edu

***Algebraic topology** to ALEJANDRO ADEM, Department of Mathematics, University of British Columbia, Room 121, 1984 Mathematics Road, Vancouver, British Columbia, Canada V6T 1Z2; email: adem@math.ubc.ca

***Combinatorics** to JOHN R. STEMBRIDGE, Department of Mathematics, University of Michigan, Ann Arbor, Michigan 48109-1109; email: FRS@umich.edu

Complex analysis and harmonic analysis to ALEXANDER NAGEL, Department of Mathematics, University of Wisconsin, 480 Lincoln Drive, Madison, WI 53706-1313; email: nagel@math.wisc.edu

***Differential geometry and global analysis** to LISA C. JEFFREY, Department of Mathematics, University of Toronto, 100 St. George St., Toronto, ON Canada M5S 3G3; email: jeffrey@math.toronto.edu

Dynamical systems and ergodic theory to AMIE WILKINSON, Department of Mathematics, Northwestern University, 2033 Sheridan Road, Evanston, IL 60208-2730; email: transactions@math.northwestern.edu

***Functional analysis and operator algebras** to MARIUS DADARLAT, Department of Mathematics, Purdue University, 150 N. University St., West Lafayette, IN 47907-2067; email: mdd@math.purdue.edu

***Geometric analysis** to TOBIAS COLDING, Courant Institute, New York University, 251 Mercer St., New York, NY 10012; email: traneditor@cims.nyu.edu

***Geometric analysis** to MLADEN BESTVINA, Department of Mathematics, University of Utah, 155 South 1400 East, JWB 233, Salt Lake City, Utah 84112-0090; email: bestvina@math.utah.edu

Harmonic analysis, representation theory, and Lie theory to ROBERT J. STANTON, Department of Mathematics, The Ohio State University, 231 West 18th Avenue, Columbus, OH 43210-1174; email: stanton@math.ohio-state.edu

***Logic** to STEFFEN LEMPP, Department of Mathematics, University of Wisconsin, 480 Lincoln Drive, Madison, Wisconsin 53706-1388; email: lempp@math.wisc.edu

***Ordinary differential equations, and applied mathematics** to PETER W. BATES, Department of Mathematics, Michigan State University, East Lansing, MI 48824-1027; email: bates@math.msu.edu

***Partial differential equations** to GUSTAVO PONCE, Department of Mathematics, South Hall, Room 6607, University of California, Santa Barbara, CA 93106; email: ponce@math.ucsb.edu

***Probability and statistics** to KRZYSZTOF BURDZY, Department of Mathematics, University of Washington, Box 354350, Seattle, Washington 98195-4350; email: burdzy@math.washington.edu

***Real analysis and partial differential equations** to DANIEL TATARU, Department of Mathematics, University of California, Berkeley, Berkeley, CA 94720; email: tataru@math.berkeley.edu

All other communications to the editors should be addressed to the Managing Editor, ROBERT GURALNICK, Department of Mathematics, University of Southern California, Los Angeles, CA 90089-1113; email: guralnic@math.usc.edu.

Titles in This Series

868 **Gelu Popescu,** Entropy and multivariable interpolation, 2006

867 **Vilmos Totik,** Metric properties of harmonic measures, 2006

866 **William Craig,** Semigroups underlying first-order logic, 2006

865 **Nathanial P. Brown,** Invariant means and finite representation theory of C^*-algebras, 2006

864 **John M. Lee,** Fredholm operators and Einstein metrics on conformally compact manifolds, 2006

863 **M. Lübke and A. Teleman,** The Universal Kobayashi-Hitchin correspondence on Hermitian manifolds, 2006

862 **Alberto Canonaco,** The Beilinson complex and canonical rings of irregular surfaces, 2006

861 **Leon A. Takhtajan and Lee-Peng Teo,** Weil-Petersson metric on the universal Teichmüller space, 2006

860 **Thomas M. Fiore,** Pseudo limits, biadjoints and pseudo algebras: Categorical foundations of conformal field theory, 2006

859 **N. Arcozzi, R. Rochberg, and E. Sawyer,** Carleson measures and interpolating sequences for Besov spaces on complex balls, 2006

858 **Enrico Valdinoci, Berardino Sciunzi, and Vasile Ovidiu Savin,** Flat level set regularity of p-Laplace phase transitions, 2006

857 **Donatella Danielli, Nocola Garofalo, and Duy-Minh Nhieu,** Non-doubling Ahlfors measures, perimeter measures, and the characterization of the trace spaces of Sobolev functions in Carnot-Carathéodory spaces, 2006

856 **Vladimir Bolotnikov and Harry Dym,** On boundary interpolation for matrix valued Schur functions, 2006

855 **Yevgenia Kashina, Yorck Sommerhäuser, and Yongchang Zhu,** On higher Frobenius-Schur indicators, 2006

854 **Noam Greenberg,** The role of true finiteness in the admissible recursively enumerable degrees, 2006

853 **Joachim Krieger,** Stability of spherically symmetric wave maps, 2006

852 **Viorel Barbu, Irena Lasiecka, and Roberto Triggiani,** Tangential boundary stabilization of Navier-Stokes equations, 2006

851 **Jie Wu,** On maps from loop suspensions to loop spaces and the shuffle relations on the Cohen groups, 2006

850 **Siegfried Echterhoff, S. Kaliszewski, John Quigg, and Iain Raeburn,** A categorical approach to imprimitivity theorems for C^*-dynamical systems, 2006

849 **Katsuhiko Kuribayashi, Mamoru Mimura, and Tetsu Nishimoto,** Twisted tensor products related to the cohomology of the classifying spaces of loop groups, 2006

848 **Bob Oliver,** Equivalences of classifying spaces completed at the prime two, 2006

847 **Eric T. Sawyer and Richard L. Wheeden,** Hölder continuity of weak solutions to subelliptic equations with rough coefficients, 2006

846 **Victor Beresnevich, Detta Dickinson, and Sanju Velani,** Measure theoretic laws for lim–sup sets, 2006

845 **Ehud Friedgut, Vojtech Rödl, Andrzej Ruciński, and Prasad V. Tetali,** A Sharp threshold for random graphs with a monochromatic triangle in every edge coloring, 2006

844 **Amadeu Delshams, Rafael de la Llave, and Tere M. Seara,** A geometric mechanism for diffusion in Hamiltonian systems overcoming the large gap problem: Heuristics and rigorous verification on a model, 2006

843 **Denis V. Osin,** Relatively hyperbolic groups: Intrinsic geometry, algebraic properties, and algorithmic problems, 2006

842 **David P. Blecher and Vrej Zarikian,** The calculus of one-sided M-ideals and multipliers in operator spaces, 2006

841 **Enrique Artal Bartolo, Pierrette Cassou-Noguès, Ignacio Luengo, and Alejandro Melle Hernández,** Quasi-ordinary power series and their zeta functions, 2005

840 **Sławomir Kołodziej,** The complex Monge-Ampère equation and pluripotential theory, 2005

839 **Mihai Ciucu,** A random tiling model for two dimensional electrostatics, 2005

838 **V. Jurdjevic,** Integrable Hamiltonian systems on complex Lie groups, 2005

837 **Joseph A. Ball and Victor Vinnikov,** Lax-Phillips scattering and conservative linear systems: A Cuntz-algebra multidimensional setting, 2005

836 **H. G. Dales and A. T.-M. Lau,** The second duals of Beurling algbras, 2005

835 **Kiyoshi Igusa,** Higher complex torsion and the framing principle, 2005

834 **Kenı́chi Ohshika,** Kleinian groups which are limits of geometrically finite groups, 2005

833 **Greg Hjorth and Alexander S. Kechris,** Rigidity theorems for actions of product groups and countable Borel equivalence relations, 2005

832 **Lee Klingler and Lawrence S. Levy,** Representation type of commutative Noetherian rings III: Global wildness and tameness, 2005

831 **K. R. Goodearl and F. Wehrung,** The complete dimension theory of partially ordered systems with equivalence and orthogonality, 2005

830 **Jason Fulman, Peter M. Neumann, and Cheryl E. Praeger,** A generating function approach to the enumeration of matrices in classical groups over finite fields, 2005

829 **S. G. Bobkov and B. Zegarlinski,** Entropy bounds and isoperimetry, 2005

828 **Joel Berman and Paweł M. Idziak,** Generative complexity in algebra, 2005

827 **Trevor A. Welsh,** Fermionic expressions for minimal model Virasoro characters, 2005

826 **Guy Métivier and Kevin Zumbrun,** Large viscous boundary layers for noncharacteristic nonlinear hyperbolic problems, 2005

825 **Yaozhong Hu,** Integral transformations and anticipative calculus for fractional Brownian motions, 2005

824 **Luen-Chau Li and Serge Parmentier,** On dynamical Poisson groupoids I, 2005

823 **Claus Mokler,** An analogue of a reductive algebraic monoid whose unit group is a Kac-Moody group, 2005

822 **Stefano Pigola, Marco Rigoli, and Alberto G. Setti,** Maximum principles on Riemannian manifolds and applications, 2005

821 **Nicole Bopp and Hubert Rubenthaler,** Local zeta functions attached to the minimal spherical series for a class of symmetric spaces, 2005

820 **Vadim A. Kaimanovich and Mikhail Lyubich,** Conformal and harmonic measures on laminations associated with rational maps, 2005

819 **F. Andreatta and E. Z. Goren,** Hilbert modular forms: Mod p and p-adic aspects, 2005

818 **Tom De Medts,** An algebraic structure for Moufang quadrangles, 2005

817 **Javier Fernández de Bobadilla,** Moduli spaces of polynomials in two variables, 2005

816 **Francis Clarke,** Necessary conditions in dynamic optimization, 2005

815 **Martin Bendersky and Donald M. Davis,** V_1-periodic homotopy groups of $SO(n)$, 2004

814 **Johannes Huebschmann,** Kähler spaces, nilpotent orbits, and singular reduction, 2004

813 **Jeff Groah and Blake Temple,** Shock-wave solutions of the Einstein equations with perfect fluid sources: Existence and consistency by a locally inertial Glimm scheme, 2004

For a complete list of titles in this series, visit the AMS Bookstore at **www.ams.org/bookstore/**.